D860. 184 .

Jeffrey Otto

INTRODUCING

GLOBAL
ISSUES

SECOND EDITION

INTRODUCING

GLOBAL
ISSUES

EDITED BY
MICHAEL T. SNARR
D. NEIL SNARR

LYNNE
RIENNER
PUBLISHERS

BOULDER
LONDON

For Ty and Ruth Shaban,
and Don and Mabel Snarr,
thanks for your love, guidance, and encouragement

Published in the United States of America in 2002 by
Lynne Rienner Publishers, Inc.
1800 30th Street, Boulder, Colorado 80301
www.rienner.com

and in the United Kingdom by
Lynne Rienner Publishers, Inc.
3 Henrietta Street, Covent Garden, London WC2E 8LU

Library of Congress Cataloging-in-Publication Data
Introducing global issues / edited by Michael T. Snarr and D. Neil Snarr. —
    2nd ed.
      p. cm.
    Includes bibliographical references and index.
    ISBN 1-58826-011-9 (alk. paper)
    1. World politics—1989–   2. International economic relations. 3. Social
history—1970. 4. Ecology. I. Snarr, Michael T. II. Snarr, Neil, 1933– .
D860 .I89   2002
909.82—dc21   2002069819

British Cataloguing in Publication Data
A Cataloguing in Publication record for this book
is available from the British Library.

Printed and bound in the United States of America

        The paper used in this publication meets the requirements
  ∞   of the American National Standard for Permanence of
        Paper for Printed Library Materials Z39.48-1984.

    5 4 3

CONTENTS

TABLES AND FIGURES

■ **TABLES**

■ FIGURES

PREFACE

In the second edition of *Introducing Global Issues,* we have updated and expanded our coverage of some of the world's most pressing problems. All of the chapters reflect significant changes in the world, including those resulting from the events of September 11, 2001; additionally, a new chapter on the global environment addresses issues of biodiversity and the global commons. In response to student suggestions, we have included a list of acronyms, along with many new tables and figures. Other ideas for improvement and general comments are welcome at michael_snarr@wilmington.edu.

* * *

We would like to express our sincere appreciation to those who made this book possible. We are first and foremost indebted to our contributors for their perseverance and hard work. Getting to know them better has been a gratifying outcome of this project. Special thanks go to the staff at Lynne Rienner Publishers—and especially to Lynne Rienner, Bridget Julian, and Lesli Brooks Athanasoulis—for their outstanding encouragement and support.

We thank our global issues students at Wilmington College for asking insightful questions and demanding a readable book, along with our colleagues, who offered many useful suggestions. Joan Skidmore and her staff provided secretarial help at every step, and Jennifer Dye and Emily Herring assisted us with research and artwork. Rodney North and Colin Frake contributed their good advice to the earlier comments of Steven L. Lamy and Jeffrey Lantis.

We owe a special debt of gratitude to our family—Melissa, Ruth, Madison, Ty, and Isaiah. Your support, love, and friendship are invaluable.

—*Michael T. Snarr*
—*D. Neil Snarr*

INTRODUCING GLOBALIZATION AND GLOBAL ISSUES

Michael T. Snarr

- Approximately 210,000 people are added to the world's population every day; that is the equivalent of 76 million people per year.
- People in more than 200 countries and territories have access to Cable News Network (CNN).
- During the 1990s, the number of undernourished people declined by approximately 40 million (UNDP 2001).
- An area of rainforest larger than a U.S. football field is destroyed every second worldwide.
- Infant mortality rate was reduced during the 1990s by over 10 percent worldwide (UNDP 2001).
- More civilians died in the twentieth century as a result of war than in the four previous centuries combined.
- Dramatic numbers of species are becoming extinct worldwide.
- More than 1 billion people live on less than one U.S. dollar per day.
- 40 million people are HIV-infected (UNAIDS 2001); approximately 11 people are infected every minute.
- Approximately 30,000 children die every day from preventable diseases.

Each of the items above is related to a global issue discussed in this book and many of them affect the reader. But what is a *global issue?* The term is used in the book to refer to two types of phenomena. First, there are those issues that are transnational, that is, they cross political boundaries (country borders). These issues affect individuals in more than one country. A clear example is air pollution produced by a factory in the United States and blown into Canada. Second, there are problems and issues that do not

necessarily cross borders but affect a large number of individuals throughout the world. Ethnic rivalries and human rights violations, for example,
may occur within a single country but have a far wider impact. Thus, global
issues either cross country borders or affect a vast number of people.

For the contributors to this volume, the primary goal is to introduce
several of the most pressing global issues and demonstrate how strongly
they are interconnected. Since these issues affect each and every one of us,
we also hope to motivate the reader to learn more about these global issues.

■ IS THE WORLD SHRINKING?

There has been a great deal of discussion in recent years about globalization, which can be defined as "the intensification of economic, political, social, and cultural relations across borders" (Holm and Sørensen 1995: 1).
Evidence of globalization is seen regularly in our daily lives. In the United
States, grocery stores and shops at the local mall are stocked with items
produced abroad. Likewise, Nike, Los Angeles Lakers, and New York Yan-
kees hats and T-shirts are easily found outside of the United States. In many
countries outside of the United States, Britney Spears, 'N Sync, and other
U.S. music groups dominate the airwaves; CNN and MTV are on television
screens; and Harry Potter is at the movies. Are we moving toward a single
global culture? In the words of Benjamin Barber, we are being influenced
by "the onrush of economic and ecological forces that demand integration
and uniformity and that mesmerize the world with fast music, fast computers, and fast food—with MTV, Macintosh, and McDonald's, pressing nations into one commercially homogeneous global network: one McWorld
tied together by technology, ecology, communication, and commerce" (Barber 1992: 53).

Technology is perhaps the most visible aspect of globalization and in
many ways its driving force. Communications technology has revolution-
ized our information systems. CNN reaches hundreds of millions of households in over 200 countries and territories throughout the world. "Computer,
television, cable, satellite, laser, fiber-optic, and microchip technologies
[are] combining to create a vast interactive communications and information network that can potentially give every person on earth access to every
other person, and make every datum, every byte, available to every set of
eyes" (Barber 1992: 58). Technology has also aided the increase in international trade and international capital flows and enhanced the spread of
Western, primarily U.S., culture.

Of course the earth is not literally shrinking, but in light of the rate at
which travel and communication speeds have increased, the world has in
a sense become smaller. Thus, many scholars assert that we are living in a

qualitatively different time, in which humans are interconnected more than ever before. "There is a distinction between the contemporary experience of change and that of earlier generations: never before has change come so rapidly . . . on such a global scale, and with such global visibility" (CGG 1995: 12).

This seemingly uncritical acceptance of the concept of globalization and a shrinking world is not without its critics. These critics point out that labor, trade, and capital moved at least as freely, if not more so, during the second half of the nineteenth century than it does now.

Second, some skeptics argue that while interdependence and technological advancement have increased in some parts of the world, this is not true in a vast majority of the South. (The terms *the South, the developing world, the less developed countries,* and *the third world* are used interchangeably throughout this book. They refer to the poorer countries, in contrast to the United States, Canada, Western Europe, Japan, Australia, and New Zealand, which are referred to as *the North, the more developed countries,* and *the first world.*) For example, Hamid Mowlana argues that "'Global' is not 'universal'" (1995: 42). Although a small number of people in the South may have access to much of the new technology and truly live in the "global village," the large majority of the population in these countries does not. Despite the rapid globalization of the Internet, it is estimated that by 2005, only one billion (or one in six people) will have access to it (UNDP 2001). Figure 1.1 further documents the lopsided nature of Internet use in the world.

In most of the poorer countries of Africa and Asia, the number of cellular mobile subscribers per 1,000 people is in single digits. In contrast, for many of the developed countries, nearly half of all people use this technology (UNDP 2001). A good example of this contrast can be seen in the current war in Afghanistan. While ultramodern U.S. jets flew above Kabul, many Northern Alliance troops were entering the city on horses and bicycles.

Similarly, one can argue that information flows, a characteristic of globalization, go primarily in one direction. Even those in the South who have access to television or radio are at a disadvantage. The globalization of communication in the less developed countries typically is a one-way proposition: The people do not control any of the information; they only receive it. It is also true that worldwide the ability to control or generate broadcasts rests in the hands of a tiny minority.

While lack of financial resources is an important impediment to globalization, there are other obstacles. Paradoxically, Benjamin Barber, who argues that we are experiencing global integration via "McDonaldization," asserts we are at the same time experiencing global disintegration. The breakup of the Soviet Union and Yugoslavia, as well as the great number of other ethnic and national conflicts (see Chapter 3), are cited as evidence of forces countering globalization. Many subnational groups (groups within

Figure 1.1 Internet Users Around the World, 1998 and 2000

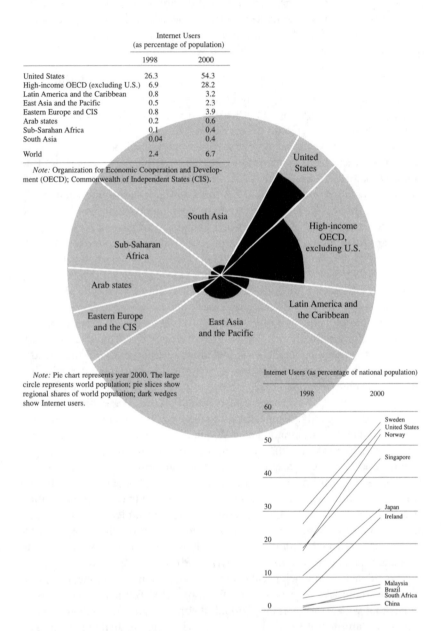

	Internet Users (as percentage of population)	
	1998	2000
United States	26.3	54.3
High-income OECD (excluding U.S.)	6.9	28.2
Latin America and the Caribbean	0.8	3.2
East Asia and the Pacific	0.5	2.3
Eastern Europe and CIS	0.8	3.9
Arab states	0.2	0.6
Sub-Sarahan Africa	0.1	0.4
South Asia	0.04	0.4
World	2.4	6.7

Note: Organization for Economic Cooperation and Development (OECD); Commonwealth of Independent States (CIS).

Note: Pie chart represents year 2000. The large circle represents world population; pie slices show regional shares of world population; dark wedges show Internet users.

Source: Adapted from UN Development Programme, *Human Development Report 2001* (New York: Oxford University Press, 2001), Feature 2.3, p. 40.

nations) desire to govern themselves; others see threats to their religious values and identity and therefore reject the secular nature of globalization. As a result, for some, globalization

> has produced not uniformity, but a yearning for a return to non-secular values. Today, there is a rebirth of revitalized fundamentalism in all the world's major religions, whether Islam, Christianity, Judaism, Shintoism, or Confucianism. At the same time the global homogeneity has reached the airwaves, these religious tenets have reemerged as defining identities. (Mowlana 1995)

None of these criticisms mean that our contemporary time period is not different in some important aspects. There is widespread agreement that communications, trade, and capital are moving at unprecedented speeds and at volumes never before seen. These criticisms do, however, provide an important caution against overstating or making broad generalizations about the process and effects of globalization.

■ IS GLOBALIZATION GOOD OR BAD?

There are some aspects of globalization that most will agree are good (for example, the spread of medical technology) or bad (for example, increased global trade in illegal drugs). Events during the war in Afghanistan in 2001–2002 revealed the dramatic contrast between friends and foes of globalization. Due to the Taliban's rejection of many aspects of Western culture, some Afghanis apparently buried their televisions and VCRs in their backyard. When Kabul was captured by the Northern Alliance it was reported that one Afghani anxiously retrieved his TV and VCR in order to view his copy of *Titanic* (Filkins 2001). Judging whether or not globalization is good is, however, complex.

The first column of Table 1.1 identifies three areas that are affected by globalization—political, economic, and cultural—and gives some examples of aspects considered positive and negative about globalization. A key aspect of political globalization is the weakened ability of the state to control both what crosses its borders and what goes on inside them. In other words, globalization can reduce the state's *sovereignty* (the state's ability to govern matters within its borders). This can be viewed as good, because undemocratic governments are finding it increasingly difficult to control the flow of information to and from prodemocracy groups. Satellite dishes, e-mail, and the World Wide Web are three examples of technology that have eroded state sovereignty. But decreased state sovereignty also means that the state has difficulty controlling the influx of illegal drugs and unwanted immigrants, including terrorists.

Table 1.1 Advantages and Disadvantages of Globalization

Effects of Globalization	Advantages	Disadvantages
Political	Weakens power of authoritarian governments	Unwanted external influence difficult to keep out
Economic	Jobs, capital, more choices for consumers	Exploitative; only benefits a few; gap between rich and poor
Cultural	Offers exposure to other cultures	Cultural imperialism

In the realm of economics, increased globalization has given consumers more choices. Also, multinational corporations are creating jobs in poor areas where people never before had such opportunities. Some critics reject these points, arguing that increased foreign investment and trade benefit only a small group of wealthy individuals and that, as a result, the gap between rich and poor grows both within countries and between countries. Related to this is the argument that many well-paying, blue-collar jobs are moving from the North to the poor countries of Latin America, Africa, and Asia.

At the cultural level, those who view increased cultural contact as positive say that it gives people more opportunities to learn about (and purchase goods from) other cultures. But critics of cultural globalization argue that the wealthy countries are guilty of cultural imperialism—that their multibillion-dollar advertising budgets are destroying the cultures of non-Western areas, as illustrated by Avon's aggressive sales strategy in the Amazon region (Byrd 1994). The fear of cultural imperialism is certainly a key component in the animosity of some Arabs toward the United States. Other critics are increasingly fearful that more and more national languages will become extinct as foreign languages, especially English, penetrate borders.

The degree to which cultural values can be "exported" is the subject of some debate. Samuel Huntington argues that

> drinking Coca-Cola does not make Russians think like Americans any more than eating sushi makes Americans think like Japanese. Throughout human history, fads and material goods have spread from one society to another without significantly altering the basic culture of the recipient society. (Huntington 1996: 28–29)

Similarly, others argue that globalization brings only superficial change. "McDonald's may be in nearly every country, but in Japan, sushi is served alongside hamburgers. In many countries, hamburgers are not even on the menu" (Mowlana 1995: 46).

It is left to the reader to determine whether globalization is having a positive or negative effect on the issues discussed in this book. Is globalization enhancing human capacity to deal with a particular issue? Or is it making it more difficult? Of course, each individual's perspective will be influenced by whether he or she evaluates these issues based on self-interest, national interest, a religious view, or from a global humanitarian viewpoint. For example, when considering the issue of free trade (Chapter 6), readers must decide how they evaluate a moral question such as, "Is free trade good or bad?" Those concerned first and foremost with self-interest will ask, How does free trade affect me? A national point of view will consider one's country and the effects of free trade on it. If readers identify with a religion, they will ask how their religion would instruct them on this question. Finally, the global humanitarian view would ask what is best for humanity in general. Therefore, readers must ask themselves what is most important when evaluating the issues discussed in this book.

■ INTERCONNECTEDNESS AMONG ISSUES

As mentioned above, a primary purpose of this book is to explore how the issues introduced in the various chapters are interconnected. In other words, the poverty chapter should not be considered separate from the chapter on population, even though these two issues are treated separately. Below are several examples of how issues discussed in this book are interconnected.

- The growth in the world's population (Chapter 9) has been dramatically affected, especially in Africa, by the AIDS crisis, which is discussed in the chapter on health (Chapter 12).
- Many of the value judgments that the trade chapter (Chapter 6) urges readers to consider are intricately linked to human rights issues (Chapter 4).
- Ethnic conflict discussed in Chapter 3 (as well as other types of conflict) often leads to internal migration as well as international population movements (Chapter 9).
- One of the recommendations for reducing poverty (Chapter 8) is to educate women and give them more decisionmaking power over their lives (Chapter 11).

The interconnectedness of these issues is even more extensive than these examples demonstrate. For instance, while an increase in AIDS will affect population growth, the connections do not end there. AIDS epidemics also lead to increased government expenditures, which can lead to increased indebtedness, which will likely lead to more poverty, and so on. Thus, each global issue discussed in this book has multiple consequences.

■ OUTLINE OF THE BOOK

This book has been organized into five parts. Part 1, which focuses on conflict and security issues, considers some of the primary sources of conflict and some of the many approaches to establishing and maintaining peace. Part 2 concentrates on economic issues ranging from international trade and capital flows to one of the major concerns that confronts humanity—poverty. Part 3 deals with issues that, although not confined to, tend to plague the poorer countries, such as population growth, health issues, and issues that affect women and children. And Part 4 focuses on environmental issues (such as global warming, ozone depletion, biodiversity, and others) and global attempts to solve them. Part 5 discusses possible future world orders, sources of hope and challenges that face us in the coming decades, and innovative actions that are being taken to make a positive impact on global issues.

■ DISCUSSION QUESTIONS

1. What examples of globalization can you identify in your life?
2. Do you think globalization will continue to increase? If so, in what areas?
3. Do you think globalization has more positive attributes or more negative attributes?
4. Can you think of additional examples of how the global issues discussed in different chapters are interconnected?

■ SUGGESTED READINGS

Barber, Benjamin R. (1996) *Jihad vs. McWorld.* New York: Ballantine Books.

Friedman, Thomas L. (2000) *The Lexus and the Olive Tree: Understanding Globalization.* New York: Anchor Books.

Hauchler, Ingomar, and Paul M. Kennedy, eds. (1994) *Global Trends: The World Almanac of Development and Peace.* New York: Continuum.

Huntington, Samuel P. (1998) *The Clash of Civilizations and the Remaking of World Order.* New York: Touchstone Books.

Jones, Ellis, et al. (2001) *The Better World Handbook: From Good Intentions to Everyday Actions.* Boulder, CO: New Society Publishers.

Kaplan, Robert D. (2000) *The Coming Anarchy: Shattering the Dreams of the Post Cold War.* New York: Random House.

Iyer, Pico (2001) *The Global Soul: Jet Lag, Shopping Malls, and the Search for Home.* New York: Vintage Books.

O'Meara, Patrick, Howard D. Mehlinger, and Matthew Krain, eds. (2000) *Globalization and the Challenges of the New Century: A Reader.* Bloomington, IN: Indiana University Press.

United Nations Development Programme (annual) *Human Development Report.* New York: Oxford University Press.

Part 1

CONFLICT AND SECURITY

2

WEAPONS PROLIFERATION AND CONFLICT

Jeffrey S. Lantis

The proliferation of weapons is one of the most serious challenges to international security today. Arms races, regional competition, and the spread of weapons technology to new countries or groups are all important dimensions of the proliferation challenge that could contribute to long-term global instability. The events of September 11, 2001, in the United States, and subsequent international developments, are a bitter reminder that the proliferation challenge is here to stay.

Proliferation is best understood as the rapid increase in the number and destructive capability of armaments. Evidence of the impact of proliferation on world affairs can be seen in the arms race between Germany and Great Britain that helped to spark World War I; the nuclear arms race between the superpowers, the United States and the Soviet Union, that brought us to the brink of a World War III; the clandestine arms buildup in Iraq that helped it fight the Gulf War; the nuclear race between India and Pakistan that culminated in a series of underground test explosions in both countries in May 1998; and the anthrax attacks in the United States in the fall of 2001.

It is important to remember, however, that proliferation is not simply a problem for politicians and military leaders. When governments choose to use weapons in conflict, they are exposing both soldiers and civilians to danger. In fact, the proliferation of weapons contributed to higher civilian casualties and greater destruction in the twentieth century than in the previous four centuries combined (Small and Singer 1982). When governments devote funds to build up large armies and weapons of mass destruction (WMD), they are also choosing to divert funds from other programs like education and health care. Clearly, citizens of the world experience direct and indirect effects of proliferation every day.

■ TYPES OF PROLIFERATION

This chapter examines four different types of weapons proliferation. As illustrated in Table 2.1, there are two broad dimensions to consider: vertical versus horizontal proliferation and conventional weapons versus weapons of mass destruction. Vertical proliferation is the development and stockpiling of armaments in one country. Horizontal proliferation is defined as the spread of weapons or weapons technology across country borders. Conventional weapons are those systems that make up the vast majority of all military arsenals—including guns, tanks, most artillery shells and bullets, planes, and ships. Weapons of mass destruction are those *special* weapons that have a devastating effect even when used in small numbers and that kill more indiscriminately than conventional weapons; they include nuclear, chemical, and biological systems.

Table 2.1   The Proliferation Matrix

	Vertical Proliferation	Horizontal Proliferation
Conventional weapons	Type I	Type II
Weapons of mass destruction	Type III	Type IV

■ *Type I: Vertical Proliferation of Conventional Weapons*

The buildup of conventional weapons arsenals in many countries is the oldest form of proliferation in human civilization and represents the core of the proliferation threat. At first glance, one might view this category of proliferation as the least threatening or most benign of all forms. Vertical conventional proliferation, however, can be a threat to international stability for at least two major reasons. First, arms buildups provide more weaponry for governments and groups to engage in more conflicts. At the same time, conventional weapons have become more sophisticated (from breech-loading rifles to precision-guided munitions) and more destructive (from mortar shells to multiple-launch rocket systems). Vertical conventional proliferation in an unregulated world market may provide determined leaders with enough incentive to order aggressive action and to actually spark conflicts. A second important danger of conventional arms buildups in one country is the social cost, which often includes serious reductions in social welfare spending by governments for citizens who can ill afford such deprivations.

More weapons mean more conflicts. Government programs to build up conventional armaments ensure that there are more weapons available for countries to engage in more conflict. Some experts believe that the simple availability of weapons systems and the development of military strategies increase the chances that a country will engage in conflict. They have argued that advances in conventional weaponry and offensive military strategies were contributing factors to the outbreaks of numerous conflicts, including both world wars and the Vietnam War. In this context, arms buildups are seen as one potential cause of war in the international system (Sagan 1986; Sivard 1991).

In traditional forms, conventional arms buildups focus on weapons systems that are considered to be most effective for the times. In the period leading up to World War I, Germany and Great Britain engaged in a race to build the most powerful and awesome warships. In the period leading up to World War II, Adolf Hitler ordered research and development of rudimentary surface-to-surface missiles and jet aircraft as a way to gain military advantage. During the Cold War, President Ronald Reagan called for the creation of a 600-ship U.S. naval fleet, with an emphasis on strong aircraft carrier battle groups and advanced submarines. More recently, attention has turned to the latest technology of warfare, including stealth planes and ships, remote-controlled aircraft, antisatellite weapons, and computer technology to give mobility and advantage to the fighting forces of the twenty-first century.

The relationship between arms buildups and the likelihood of conflict is multiplied by the fact that conventional weapons have become more sophisticated and destructive over the years. "Smart" conventional bombs and precision-guided munitions have improved in terms of both accuracy and destructive capability. Shoulder-launched missiles empower one person with the ability to shoot down large aircraft. The modernization of conventional weapons such as fuel-air explosives and the faster and more accurate M1A1 tank also poses a great threat to soldiers and civilians.

Finally, it is important to remember that conventional arms have been used repeatedly in conflict since the end of World War II. From landmines to fighter jets, conventional weapons have been blamed for roughly 50 million deaths around the globe since 1945. Individuals, groups, and governments have all built and used conventional weapons to achieve their goals.

The social costs of arms buildups. At the beginning of the twenty-first century, many governments have sizeable conventional arsenals. U.S. military expenditures have topped $250 billion annually in the past two decades, with the majority of these funds going to support high levels of conventional weaponry and troops. In 2001, the Bush administration sought a total defense budget of $305 billion to support an active-duty military strength of 1,370,000 soldiers and a force structure composed of ten army divisions, twelve navy aircraft carrier task groups, three Marine Corps divisions, five army special forces groups, and thirteen air force combat wings (Cohen

2001). More than 250,000 of these U.S. soldiers were stationed abroad (with large deployments in Europe, the Persian Gulf region, and Asia). Critics point out that in relative terms, U.S. defense expenditures in 2001 were more than six times larger than that of its nearest potential competitor, and more than twenty-three times larger than the combined spending of the seven "rogue states," or countries traditionally identified as the most likely adversaries of the United States: Cuba, Iran, Iraq, Libya, North Korea, Sudan, and Syria (Center for Defense Information 2001). Table 2.2 illustrates the broader context of changing levels of global and regional defense expenditures in the post–Cold War era.

Though Table 2.2 shows a slight decline in global defense spending in the early to mid-1990s, levels are once again on the rise. In 2000, global military expenditures reached $798 billion, or $130 for every person on the planet. Military expenditures rose in all parts of the world between 1998 and 2000, but African countries saw the sharpest increase of defense spending, 37 percent in that two-year period (SIPRI 2001). It is clear that countries continue to spend hundreds of billions of dollars every year on the military. This has led many critics to charge that there are dangerous social costs in the trade-off between "guns and butter," and the end of the Cold War drew new attention to this difficult balance between military and social spending.

In 1990, the United States ranked first in the world in terms of military spending but compared rather poorly with other countries on various social indicators. To some degree this was a function of priorities in government spending during the Cold War. Table 2.3 shows how the United

Table 2.2 World and Regional Military Expenditures, 1991, 1995, and 2000 (in U.S.$ billions at constant 1998 values)

Selected Regions	1991	1995	2000	Percentage of Change, 1991–2000
Africa	11.6	10.1	13.8	+20
Asia	97.9	112	123	+26
Central America	2.2	2.7	3.0	+29
Europe	302	239	240	−19
Middle East	70.7	47.9	60.9	−14
North America	345	307	288	−16
South America	16.5	22.9	26.3	+59
World	860	742	756	−12

Source: SIPRI (Stockholm International Peace Research Institute), SIPRI Yearbook 2001 (London: Oxford University Press, 2001).
Notes: Figures may not always add up to totals because of conventions of rounding and estimates of expenditures. Data for some countries have been excluded because of lack of information. Africa figures do not include expenditures of Congo, Libya, and Somalia; Asia excludes Afghanistan; Europe excludes Yugoslavia; and Middle East excludes Iraq.

Table 2.3 The Social Costs of U.S. Military Expenditure During the Cold War, 1990

Military expenditure	1
Literacy rate	4
Gross national product per capita	6
Per capita expenditure for education	9
Life expectancy	10
Average scores of students on science and math tests	13
Infant mortality rate	21
Population per physician	22

Source: Ruth Leger Sivard, *World Military and Social Expenditures 1993* (Washington, DC: World Priorities, 1993).
Note: Rank compared with 140 other countries.

States compared with other countries in terms of social welfare standards, literacy rates, sanitation, education, and health care during the Cold War. A related study found that when military expenditures rose in developing countries, the rate of economic growth declined and government debt increased (Nincic 1982). This has led some to conclude the sad truth that many countries became more concerned with defending their citizens from foreign attack in the twentieth century than they were with protecting them from social, educational, and health insecurities at home.

■ Type II: Horizontal Proliferation of Conventional Weapons

A second category of proliferation is the horizontal spread of conventional weapons and related technology across country borders. The main route of the spread of conventional weaponry is through legitimate arms sales. But the conventional arms trade has become quite lucrative, and many experts are concerned that the imperative of the bottom dollar is driving us more rapidly toward global instability.

Arms dealers. The conventional arms trade has become a very big business, and seven powerful countries—the United States, France, Russia, Great Britain, China, Germany, and Israel—are responsible for more than 90 percent of global sales. Patterns in the arms trade have changed over time, however. In 1987, the Soviet Union was at the top of the arms trade, dominating the market with 46 percent of all sales. But as Soviet and (later) Russian sales levels plummeted, the United States quickly emerged as the new leader. In 2000, U.S. companies exported at least $26.5 billion worth of conventional arms, or 49 percent of the global market share. Their nearest competitor, France, exported $9.8 billion in arms (18 percent of the global market); Great Britain exported $8.9 billion in arms (16 percent). Other major exporters controlled a significantly smaller percentage of the global trade, including Russia with $2.8 billion in sales, Israel with $1.3 billion, Germany at $0.8 billion, and China at $0.5 billion in global arms sales (IISS 2000).

Arms customers. Who are the primary buyers of all of these weapons? Generally speaking, U.S. defense contractors have sold most hardware to allied countries. For example, in 1998, traditional allies like Germany purchased $259 million in armaments from U.S. defense contractors; the government of Greece bought $531 million in U.S. arms. In the Middle East, U.S. defense contractors sold $2.3 billion worth of armaments to Saudi Arabia, $1 billion to Egypt, and $628 million to Israel (which included armored combat vehicles and advanced fighter jets). Bahrain purchased $286 million in U.S. military hardware, and Turkey spent $241 million. In Asia, allies purchased large numbers of U.S. conventional weapons systems in 1998. This included sales to Taiwan ($441 million), Japan ($348 million), and South Korea ($267 million).

Arms sales are not always made to countries considered traditional allies, however. From 1984 to 1989, the People's Republic of China spent some $424 million on U.S. weapons, and these arms deals were stopped only after the Tiananmen Square massacre of prodemocracy activists in the summer of 1989. Through legitimate means, Iraqi president Saddam Hussein purchased a massive conventional arsenal on the international arms market. In 1990, estimates of the arsenal included a total of 5,500 tanks; 4,000 pieces of heavy artillery; 7,500 armored personnel carriers; and 700 planes. Arms sales to Iraq by friends and allies came back to haunt the United States, however, during the Gulf War. Indeed, the sale of conventional weapons raises real concern about the potential for "deadly returns" on U.S. investments (Laurance 1992).

Concerned world citizens have devoted new attention in the post–Cold War era to the role of conventional arms trades in fueling civil wars and violence in developing countries. Legal and illegal arms transfers have contributed to civil wars around the world in the past decade, including conflicts in Mozambique, Bosnia, Sierra Leone, Afghanistan, Algeria, Sudan, Colombia, Sri Lanka, Chechnya, Congo, and other developing countries. Sadly, most of the casualties in these conflicts have been civilians, and children have been especially victimized. According to the United Nations Development Programme, more than 2 million children were killed and another 4.5 million disabled in civil wars and conflicts between 1987 and 1997 (Klare 1999).

■ *Type III: Vertical Proliferation of*
Weapons of Mass Destruction

The vertical proliferation of weapons of mass destruction is another serious threat to international security. There are several important dimensions of this problem, including the range and variety of modern WMD systems, incentives for states to build nuclear weapons, and the patterns of vertical WMD proliferation.

Types of weapons of mass destruction. There are three different types of weapons of mass destruction: nuclear, biological, and chemical. These are often examined as a group, but it is important to note that their effects and their potential military applications are quite different.

Nuclear fission was discovered in 1938, and scientists like Albert Einstein soon called on governments to sponsor an exploration of its potential. Atomic weapons were first developed by the U.S. government through the five-year, $2 billion secret research program during World War II known as the Manhattan Project. On August 6, 1945, the United States dropped a 12.5-kiloton atomic bomb on Hiroshima, Japan. This weapon produced an explosive blast equal to that of 12,500 tons of conventional high explosives (such as TNT) and caused high-pressure waves, flying debris, extreme heat, and radioactive fallout. A second bomb was dropped on Nagasaki on August 9, 1945, and the Japanese government surrendered one day later (Schlesinger 1993).

The use of atomic bombs to end World War II in 1945 was actually the beginning of a very dangerous period of spiraling arms races between the United States and the Soviet Union. The Soviet regime immediately stepped up its atomic research and development program. In 1949, it detonated its first atomic test device and joined the nuclear club. By the 1980s, the Soviet Union had accumulated an estimated 27,000 nuclear weapons in its stockpile. Both superpowers also put an emphasis on diversification of their weapons systems. The symbolic centerpiece of each side's nuclear arsenals was their land-based inter-continental ballistic missiles (ICBMs), capable of accurate attacks when launched thousands of miles to their targets. But each side had also deployed nuclear weapons on submarines; in long-range bombers; as warheads on short-range, battlefield missile systems; and even in artillery shells and landmines.

Chemical weapons and biological weapons. Chemical weapons, another class of weapons of mass destruction, work by spreading poisons that can incapacitate, injure, or kill through their toxic effects on the body. These antipersonnel weapons can be lethal when vaporized and inhaled in very small amounts or when absorbed into the bloodstream through skin contact. Examples of chemical weapons range from mustard gas used during World War I to nerve agents such as Sarin (employed by a radical religious cult in Japan to kill, injure, and terrorize civilians in Tokyo in 1995). Nerve agents are invisible and odorless, and they can produce a loss of muscle control and death within minutes for untreated victims.

Many governments and independent groups have funded chemical weapons research and development programs. In fact, chemical weapons are relatively simple and cheap to produce compared with other classes of WMDs. Any group with access to chemical manufacturing plants can develop variants of commonly used, safe chemicals to create dangerous weapons of mass destruction. The first recorded use of chemical weapons in warfare occurred in the fifth century B.C.E. when Athenian soldiers poisoned their enemy's water supply with a chemical to make them sick. During World War I, more than 120,000 tons of chemical weapons were used by both sides on the western front. In 1917 alone, mustard gas attacks killed 91,000 soldiers and injured more than one million. The last known widescale use occurred during the Iran-Iraq War (1980–1988), where an estimated 13,000 soldiers were killed by chemical agents (McNaugher 1990).

As dangerous as chemical weapons can be, biological agents are much more lethal and destructive. Biological agents are basically disease-causing microorganisms such as bacteria, viruses, or fungi that cause illness or kill the intended target after an incubation period (if left untreated). A more lethal derivative of biological weapons—toxins—can cause incapacitation or death within minutes or hours. Examples include anthrax, which was used in terrorist attacks in the United States in fall 2001. Anthrax is a disease-causing bacteria that contains as many as 10 million lethal doses per gram, but the key to its lethality lies in the effectiveness of delivery systems and also, of course, in medical treatment of infected victims.

Like chemical agents, biological and toxic weapons are relatively easy to construct and have a high potential lethality rate. Any government or group with access to pharmaceutical manufacturing facilities or biological research facilities can develop biological weapons. And, like the other classes of WMD systems, information about the construction of such systems is available in the open scientific literature and on the Internet.

Why build WMD systems? There are two basic reasons why countries build weapons of mass destruction: security and prestige. First, many government leaders genuinely believe that their state security is at risk without such systems. During the Cold War, the United States and Soviet Union established large nuclear weapons stockpiles—but they also developed sizeable arsenals of chemical and biological weapons. U.S. policy on bioweapons development was finally reversed by President Richard Nixon in the early 1970s, while clandestine Soviet research and development continued to the end of the Cold War.

The standoff between India and Pakistan is another prime example of the drive for WMD security. After years of rivalry and border skirmishes between the countries, India began a secret program to construct an atomic device that might swing the balance of regional power in its favor. In 1974, the Indian government detonated what it termed a "peaceful nuclear explosion"—

signaling its capabilities to the world and threatening Pakistani security. For the next twenty-five years, both Pakistan and India secretly developed nuclear weapons in a regional arms race. In May 1998, the Indian government detonated five more underground nuclear explosions, and the Pakistani government responded to the perceived threat with six nuclear explosions of its own. The two governments acknowledged their nuclear capabilities to the world, and relations between the neighbors have been quite tense. Another example comes from the Middle East, where Israel is suspected of having developed dozens of nuclear devices for potential use in its own defense. Recently revealed diplomatic cables suggest that the Israeli government secretly threatened to use these systems against Iraq during the Gulf War if Israel came under chemical or biological weapons attack (Schlesinger 1993).

Second, some governments have undertaken WMD research and development programs for reasons of prestige, national pride, or influence. It became clear to some during the Cold War that the possession of WMD systems lent a certain level of prestige, power, and even influence to state affairs. At a minimum, the possession of WMD systems—or a spirited drive to attain them—would gain attention for a country or leader. North Korea's drive to build a nuclear device based on an advanced uranium enrichment process drew the attention of the United States and other Western powers in the early 1990s. After extensive negotiations, North Korea was offered new nuclear energy reactors in exchange for a promise not to divert nuclear material for a bomb program.

Other government leaders pursue the development of WMD arsenals because they believe that it will help them gain political dominance in their region. To illustrate this dynamic, Gerald Steinberg (1994), an expert on proliferation, relates the story of clandestine Iraqi government efforts to develop a WMD arsenal. Iraqi leader Saddam Hussein ordered the creation of a secret WMD research and development program and began to acquire nuclear technology and materials from France, Germany, the United States, and other countries in the late 1970s. While research scientists in the program worked on uranium enrichment, Saddam Hussein worked to strengthen his political profile in the region and to improve relations with key Arab states. Meanwhile, the Israeli government tried to stop the clandestine nuclear program by carrying out a devastating air strike against Iraq's nuclear research reactor at Osiraq in 1981. But the determined Iraqi drive for regional influence was really only stopped by the efforts of the U.S.-led international coalition in the Gulf War and the dispatch of a United Nations (UN) special commission to investigate and dismantle the Iraqi WMD development program. Broadly speaking, Iraqi proliferation efforts were part of a larger scheme to gain prestige, power, and influence in the Middle East.

Type IV: Horizontal Proliferation of Weapons of Mass Destruction

The horizontal proliferation of WMD systems represents the final dimension of this challenge to international peace and stability. In fact, the spread of these weapons and vital technology across state borders is often viewed as the most serious of all proliferation threats.

Nuclear arsenals. The massive buildup of nuclear arsenals by the superpowers was not the only game in town during the Cold War. In fact, while the Soviet Union and United States were stockpiling their weapons, other countries were working to join the nuclear club through both open and clandestine routes.

Today, the United States, Russia, France, Great Britain, China, India, and Pakistan all openly acknowledge possessing stocks of nuclear weapons (Figure 2.1). At the height of the Cold War, the United States and Soviet Union supported key allies by secretly authorizing the transfer of sensitive nuclear weapons technology to other research and development programs. In 1952, Great Britain successfully tested an atomic device and eventually built a nuclear arsenal that today numbers about 200 weapons. France officially joined the nuclear club in 1960 and built a somewhat larger nuclear arsenal of an estimated 420 weapons. The People's Republic of China detonated its first atomic device in 1964 and built an arsenal of about 300 nuclear weapons during the Cold War (McGwire 1994).

The controlled spread of nuclear weapons and weapons technology from the superpowers to key allies is not the only route by which countries might obtain valuable information and materials. Several less developed countries began secret atomic weapons research and development projects after World War II. As noted earlier, states like India, Pakistan, and Israel pursued clandestine WMD programs because of concerns about security and prestige. In some cases, these efforts were facilitated by covert shipments of material and technology from the great powers, but research and development of WMD systems was also aided by the availability of information in the open scientific literature (and by the resourcefulness of scientists and engineers).

When the Indian government detonated its first nuclear explosion in 1974, it symbolically ended the monopoly on nuclear systems held by the great powers. India actually obtained nuclear material for its bomb by diverting it from a Canadian-supplied nuclear energy reactor that had key components originally made in the United States. Most experts believe that India now possesses a significant stockpile of about fifty unassembled nuclear weapons. The 1974 Indian detonation was, of course, a catalyst for the Pakistani government to step up its research and development program, and

Figure 2.1   Nuclear Status Around the World, 1998

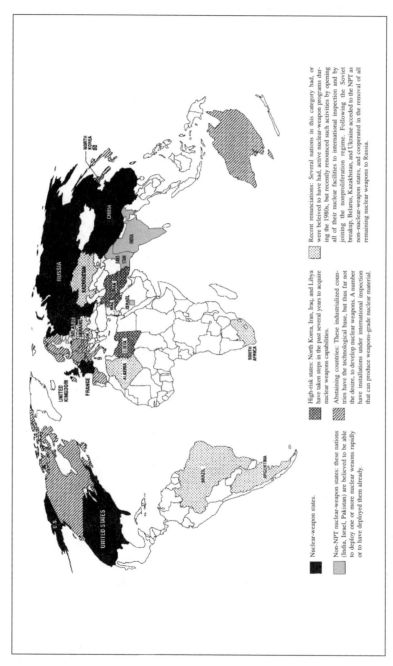

Source: Rodney W. Jones, Mark G. McDonough, Toby F. Dalton, and Gregory D. Koblentz, *Tracking Nuclear Proliferation: A Guide in Maps and Charts* (Washington, DC: Carnegie Endowment for International Peace, 1998).

today most experts believe that Pakistan has an arsenal of dozens of weapons that could be quickly assembled for use. The test explosions sponsored by both governments in 1998 put the world on notice regarding their capabilities. Finally, Israel may possess as many as 100 nuclear weapons. The Israeli nuclear program was a derivative of research and development projects in the United States and, ironically, the Soviet Union. Like India, the Israeli government proved to be quite resourceful in adapting existing technologies to construct its arsenal (Forsberg, Driscoll, Webb, and Dean 1995).

According to the Carnegie Endowment for International Peace, several other countries are considered "high-risk" proliferants. Iran, Iraq, Libya, and North Korea have all attempted to develop or acquire nuclear weapons, but they are limited by international supply controls and sanctions. While the government of North Korea signed a framework agreement in 1994 in which it pledged not to develop nuclear weapons, experts believe that North Korea may already have sufficient weapons material for one or more nuclear devices. The governments of Iran and Libya have attempted to develop nuclear weapons, but experts believe that their research and development facilities are quite limited to date. Finally, Saddam Hussein's Iraqi regime was well on the way to the development of nuclear weapons prior to the Gulf War. From 1991 to 1998, Iraq's research and development program was dismantled and monitored by the United Nations Special Commission (Jones, McDonough, Dalton, and Koblentz 1998), but Western governments remain concerned about the status of clandestine weapons programs in Iraq today.

Finally, there are former nuclear states that have made political decisions to give up their weapons and end research and development efforts. Included in this group are South Africa, Brazil, Argentina, and three former Soviet republics: Belarus, Kazakhstan, and Ukraine. The South African government admitted that it had constructed six nuclear devices for self-defense in the 1970s and 1980s. But the government decided to destroy these weapons in 1990—unilaterally removing themselves from the nuclear club. Argentina and Brazil renounced their past efforts to develop nuclear arsenals and have developed a new antinuclear profile for the region. The three former Soviet republics had about 3,000 strategic nuclear weapons stationed on their territory after the breakup of the Soviet Union. Soon after gaining their independence, however, the three republics agreed to become non–nuclear weapons states under the Treaty on the Non-Proliferation of Nuclear Weapons (or Non-Proliferation Treaty, NPT). In 1992, they signed on to the Lisbon Protocol to the Strategic Arms Reduction Treaty (START), agreeing to transfer all the nuclear warheads on their territory to Russia in exchange for economic assistance from the United States (McGwire 1994).

The spread of chemical and biological weapons. At least twenty countries are suspected to possess chemical and biological weapons, and many have chosen to develop these weapons through clandestine routes. Table 2.4 illustrates the range of actors involved in the proliferation of weapons of mass destruction.

WMD terrorism. The horizontal proliferation of WMD systems raises another concern about international security at the dawn of the twenty-first century: the use of nuclear, chemical, or biological weapons systems in terrorist attacks. Unfortunately, this fear was realized in 2001 with a wave of "toxic terrorism" in the United States. These attacks caused panic in many industrialized countries and increased public concerns about other, potential threats such as the use of smallpox or the plague as a terrorist weapon.

A recent study of documented terrorist incidents involving chemical and biological agents reveals that there were 101 cases of such attacks worldwide between 1900 and 1999. According to the database compiled by

Table 2.4 Countries Suspected to Possess Chemical or Biological Weapons Capabilities

Chemical Weapons	Biological Weapons
Bulgaria	China
China	Iran
Egypt	Iraq[a]
France	Israel
India	Libya
Iran	North Korea
Iraq[a]	Syria
Israel	Taiwan
Libya	
Myanmar (Burma)	
North Korea	
Pakistan	
Russia	
Saudi Arabia	
South Korea	
Syria	
Taiwan	
United Kingdom	
United States	
Vietnam	

Sources: E. J. Hogendoorn, "A Chemical Weapons Atlas," *Bulletin of Atomic Scientists* (September–October 1997): 35–39; "Chemical Weapons in the Middle East," *Arms Control Today* (October 1992): 44–45; Randall Forsberg, William Driscoll, Gregory Webb, and Jonathan Dean, *The Nonproliferation Primer: Preventing the Spread of Nuclear, Chemical, and Biological Weapons* (Cambridge, MA: Institute for Defense and Disarmament Studies, 1995).

Note: a. Since the end of the Gulf War, the United Nations Special Commission (UNSCOM) has destroyed more than 4.8 million liters of Iraqi chemical agents and 1.8 million liters of precursor chemicals.

the Monterey Institute's Center for Nonproliferation Studies, chemical or biological attacks have accounted for a total of 103 fatalities and 5,554 injuries in the past century. Although two-thirds of all documented incidents occurred outside the United States, some of the more infamous attacks were closer to home. In 1984, for example, followers of the Rajneeshee cult contaminated restaurant salad bars in The Dalles, Oregon, with salmonella bacteria, sickening 750 people for several days. Somewhat surprisingly, the attack was intended not to kill innocent victims but rather to keep voters at home in order to influence the outcome of a local election (Tucker 2000).

Many experts believe that the horizontal spread of nuclear weapons, material, and know-how has increased the likelihood that a group or state will attempt an act of nuclear terrorism in the future. This is of particular concern given the chaos and instability surrounding the nuclear arsenal of the former Soviet Union, and there have been numerous reports of attempts to buy or steal nuclear warheads in that region. In January 1991, armed Azeri rebels in Azerbaijan penetrated a Soviet base on which tactical nuclear weapons were stored and gained physical access to a nuclear warhead for a short period before being ejected by Soviet troops. Later that year, a nuclear expert from Greenpeace conducted a secret investigation to see how easily one might buy a Soviet nuclear device on the black market. With very little effort he found a group of disgruntled Russian soldiers willing to sell a Soviet nuclear warhead for $250,000, but he quickly reneged on the deal and revealed this news to international regulatory authorities (Hersh 1994).

The emergence of fifteen newly independent states in the region with very porous borders also increased the likelihood that nuclear materials and know-how would be smuggled out of the country. In Germany, for example, hundreds of arrests were made in connection with attempts to smuggle nuclear materials out of the former Soviet Union in the early 1990s. In October 1992, two containers of radioactive material were discovered by the police in Frankfurt containing amounts of cesium and strontium misappropriated from scientific or medical establishments in the former Soviet Union. While these materials would not be sufficient to construct a traditional nuclear device, experts highlight such cases to show that the construction of "radiological devices" (smaller quantities of radioactive material used with conventional explosives) would be much more plausible. Even low-grade radioactive material such as cesium-137, commonly used in hospital x-rays, would be useful as a basis for a radiological terrorist device. Meanwhile, nuclear scientists in the former Soviet republics (who earn an average monthly salary of $30 in defunct research facilities) may have been lured to less developed states like Iran to work in budding nuclear research programs by the promise of high wages and social status. All of these factors suggest that the threat of WMD terrorism has indeed increased in the post–Cold War era, although experts are quick to add that there have been only a few, isolated cases in which attempts to smuggle significant weapons-grade material have been identified (Forsberg, Driscoll, Webb, and Dean 1995).

■ GLOBAL SOLUTIONS?

■ The Nonproliferation Regime

Proliferation is a very complex and multilayered challenge to international security. Many world leaders and experts have recognized this threat and have addressed the proliferation challenge through a series of regional and global initiatives. The scope, number, and momentum of these nonproliferation initiatives have steadily increased over the past few decades, but it should be noted that many skeptics question the effectiveness of these efforts in the post–Cold War era.

In some ways, the global nuclear nonproliferation movement began even before the first use of atomic weapons in 1945. Politicians, military leaders, and scientists involved in the Manhattan Project recognized from the outset that such weapons were special and more dangerous than other systems. President Harry Truman, who had ordered the use of atomic bombs over Hiroshima and Nagasaki, authorized his ambassador to the United Nations, Bernard Baruch, to deliver a proposal to the organization calling for all nuclear materials and technology to be placed under UN oversight. While the plan did not receive widespread support, it demonstrated a first step toward global consideration of proliferation problems and set the stage for later progress on the issue.

In the 1960s, world leaders agreed to new initiatives, including the Partial Nuclear Test Ban Treaty. This agreement banned nuclear tests in the atmosphere, in outer space, and under water, and it was originally signed by leaders from the Soviet Union, Great Britain, and the United States. In 1967, the Treaty for the Prohibition of Nuclear Weapons in Latin America created a large nuclear-free zone. Signatories to this treaty pledged to use nuclear facilities only for peaceful purposes. Most important, the treaty outlawed the testing or acquisition of nuclear weapons in the region and even precluded third parties from bringing weapons to the region (Davis 1991).

The Non-Proliferation Treaty. The NPT represented one of the most significant advances in the development of the global nonproliferation regime. The NPT was an agreement to halt the spread of nuclear weapons beyond the five declared nuclear powers. First opened for signature in 1968, the treaty had ambitious goals for both vertical and horizontal proliferation. Article I of the treaty dictated that no nuclear weapons state (defined by the treaty as a state that detonated a nuclear explosive prior to 1967) would transfer "directly or indirectly" nuclear weapons, explosive devices, or control over these weapons to another party. Article II stipulated that no non–nuclear weapons state could receive, manufacture, obtain assistance for manufacturing, or otherwise try to acquire nuclear weaponry. Another significant element of the treaty was Article VI, which obligated all nuclear states to pursue the ultimate goal of eliminating nuclear weapons and completing a treaty on general disarmament under strict and effective international control. In many ways, the NPT represented the crowning achievement of global nonproliferation efforts during the Cold War.

Related nonproliferation initiatives. Several other significant agreements have followed in the spirit of the NPT. The Biological Weapons Convention of 1972 was the first formal effort to gain some control over the world's deadly biological arsenal. More than 140 countries have agreed to ban "the development, production and stockpiling of microbial or other biological agents." The convention, however, did not sanction nonsignatories and did not preclude research on biological weapons (Davis 1991). In the same spirit, world leaders engaged in years of negotiation to draft a Chemical Weapons Convention (CWC), which was opened for signature in January 1993. This treaty committed all signatories to eliminate their stockpiles of chemical weaponry and to halt all development efforts. In addition, it included a set of verification procedures somewhat more stringent than those under the NPT. These procedures supported the rights of a new CWC inspectorate to conduct rigorous investigations and surprise "challenge inspections" of suspected chemical weapons programs in signatory states. Ratification by the legislatures of sixty-five countries in the system was required for the CWC to come into force. After a great deal of debate about the implications of the treaty for U.S. national security and sovereignty, the U.S. Congress finally ratified the agreement in 1997.

In 1972, the United States and the Soviet Union negotiated their first two major arms control treaties, the beginning of an important period of improved relations between the superpowers once locked in a bitter Cold War. The Strategic Arms Limitation Treaty (SALT I) called for limits on the number of nuclear launch platforms including missiles and strategic bombers. The Anti-Ballistic Missile (ABM) Treaty limited each superpower to the deployment of defensive, ground-based antimissile systems at only two sites (later reduced to one). According to Article 5 of the treaty, each party pledged "not to develop, test, or deploy ABM systems or components which are sea-based, space-based, or mobile land-based" beyond allowed limits. Although the ABM Treaty was of unlimited duration, either side had the right to withdraw from the agreement by giving six-months notice. Both the SALT and ABM treaties paved the way for further bilateral agreements to control and reduce the numbers of nuclear weapons in the 1970s and 1980s.

In 1996, the Comprehensive Nuclear Test Ban Treaty (CTBT), another nonproliferation initiative, was opened for signature. A large majority of UN member states voted to support the CTBT (a treaty that would eliminate all actual nuclear testing), but to become international law the treaty required the signature of all forty-four countries known to possess nuclear reactors. By early 1998, representatives of the five declared nuclear powers had all signed the CTBT and delivered the treaty for ratification by their legislatures.

A series of problems has blocked further progress on the CTBT, however. The governments of both India and Pakistan have refused to sign the treaty. India has claimed that it wants the CTBT to be stronger in order to force nuclear states' compliance with Article VI of the NPT, and the Pakistani government has stated that it would not sign the CTBT without Indian cooperation. Their nuclear tests of May 1998 underscored their resistance to this latest initiative of the nonproliferation regime. Several other states also remain reluctant to sign the treaty, including Libya, Cuba, and Syria. In October 1999, the U.S. Senate engaged in heated debates over the ratification of the CTBT, verification procedures, problems with nonsignatories, and the national security implications of the agreement. In a move that stunned the world, the Senate rejected the treaty, and the Bush administration has shown no inclination to renew consideration of the CTBT.

■ Controlling Weapons at the Point of Supply

Given serious concerns about the implications of the spread of WMD technology around the world, governments have also devoted attention to controlling weapons transfers at the point of supply. The NPT, for example, did not prevent states from exporting "peaceful" nuclear energy reactors and other types of materials that could potentially be adapted for use in the development of WMD programs. To address such concerns, major supplier states—including the Soviet Union, Japan, France, the United States, Great Britain, West Germany, and Canada—agreed in 1976 to establish a "trigger list" of items that could be sold to other countries only under stringent safeguards. Representatives of these states met in London, and this "London Club" established and coordinated a supply control group. In the 1980s, supplier states established the Missile Technology Control Regime, which prohibited the transfer of essential technology for the development of ballistic missile systems.

Like the NPT, however, supply control efforts have had only a mixed rate of success. They helped limit missile development projects under way in South America and the Middle East but allowed some twenty countries to join the ballistic missile club. These supply controls did not prevent Iraq from making significant progress toward the development of nuclear weapons through the modification of civilian scientific technology that was adapted for military use. And they did not prevent Iraq from manufacturing and modifying the Scud-B missiles that were used against Israel and Saudi Arabia during the Gulf War—and that were capable of carrying chemical warheads. Meanwhile, Pakistan developed its own ballistic missile, the Hatf, and acquired about thirty nuclear-capable medium-range M-11 missiles from China (McNaugher 1990). In 2002 the North Korean government continues to develop a long-range Taepo Dong missile that may someday have the potential to reach the continental United States.

On the conventional weapons front, there is growing recognition that arms transfers—even "small arms" like guns and antitank weapons—represent a fundamental threat to international security. Security experts and government leaders have been discussing ways to increase the transparency of the conventional arms trade by making more information available on arms transfer policies and data. In 2001, the United Nations estimated that there were more than 500 million small arms in the world, 40–60 percent of which have been acquired illegally through black markets (Crossette 2001; Klare 1999). The small arms trade fuels civil wars in developing countries, empowers organized crime, and enables terrible human rights violations. Taliban control of Afghanistan, for example, was maintained largely by a combination of the force of will, fundamentalist religious beliefs, and small arms.

In summer 2001, the United Nations sponsored its first conference on the small arms trade, with the expressed intention of controlling weapons at the point of supply. Diplomats at the conference spoke of heartfelt concern about the weapons trade and its impact on the lives of civilians in war-torn developing countries. Representatives from 170 nations reached consensus on a treaty to combat illegal small arms trafficking on July 21, 2001, but many returned home from the summit with mixed feelings about the final agreement. During the deliberations the United States had threatened to block international consensus on the treaty if language to restrict civilian gun ownership and limit U.S. government efforts to supply small arms to nonstate actors had been left in the deal. The governments of China, India, and Russia also expressed serious reservations about limitations on arms transfers. After eleventh-hour negotiations, diplomats agreed to a watered-down version of the small arms trade treaty that would make some important statements about this area of proliferation.

■ The Missile Defense Controversy

U.S. government leaders have expressed concerns about weapons proliferation for decades, but presidents have developed markedly different approaches to the problem over time. President Truman, troubled by the implications of the decision to drop two atomic bombs on Japan at the end of

World War II, supported the Baruch Plan for the transfer of nuclear technology to UN control. President John F. Kennedy helped draft the Partial Nuclear Test Ban Treaty with the Soviet Union soon after the Cuban Missile Crisis. President Richard Nixon led U.S. negotiations with the Soviet Union that established the first major arms control treaties between the superpowers, SALT I and the ABM Treaty. In 1977, President Jimmy Carter proclaimed that one of his administration's goals would be the elimination of nuclear weapons from the face of the earth.

Ronald Reagan, elected president in 1980, brought a new perspective to the challenge of proliferation. He was quite concerned by the fragility of nuclear deterrence with the Soviet Union (which he termed the "Evil Empire") and consulted with advisers about alternative strategies. Drawing on scientific and military advice, Reagan announced in March 1983 that he was ordering "a comprehensive and intensive effort to define a long-term research and development program to begin to achieve our ultimate goal of eliminating the threat posed by strategic nuclear missiles"(Michalak 2001). Specifically, the government would fund an intensive program to develop a space-based missile defense shield, known as the Strategic Defense Initiative (SDI, or as it commonly became known, "Star Wars"). While there was considerable debate about the scientific feasibility of SDI, critics were most vocal about the legal implications of challenging the ABM Treaty with the Soviet Union. Attention to SDI research was diverted, however, by the end of the Cold War and the disintegration of the Soviet Union. In the 1990s, the Clinton administration devoted few resources and little enthusiasm to research and development of such a system.

Yet the issue was not dead. During the 2000 presidential election campaign, Republican candidate George W. Bush proclaimed that the development of a high-technology national missile defense (NMD) system would be a centerpiece of his national security policy platform. This quickly became a controversial new dimension of the proliferation challenge in 2001. Those who supported a renewed U.S. effort to develop an NMD system believed that the United States faced new threats of missile attack—not from Russia or other former Soviet republics—but more likely from the actions of a terrorist group, a "rogue state" like North Korea, or an accidental launch by larger nuclear powers. In this sense, NMD was a response to the contemporary proliferation challenge. Supporters argued that a missile defense shield would provide protection as well as deterrence against the possibility of attack. The plan to develop land-, sea-, and space-based systems would have significant implications for U.S. security in the twenty-first century.

Furthermore, President George W. Bush and key administration officials contended that the ABM Treaty would not, and should not, restrict U.S. research and development initiatives. They argued that the 1972 treaty was outmoded because the Soviet Union and the Cold War had passed into history and were supplanted by more diverse and unpredictable threats from terrorist groups and rogue nations. NMD would be designed purely as a defensive shield, they claimed, and should therefore not concern foreign governments.

However, U.S. plans to develop NMD drew heated opposition both at home and abroad. Critics launched a series of challenges to the initiative including questions about the scientific feasibility of the program, the threat of violation of the ABM Treaty, and destabilization of relations between the United States and other major nuclear powers Russia and China (Newhouse 2001). The U.S. military conducted several failed tests of an interceptor rocket system designed to shoot down a dummy missile in flight, raising questions about the scientific plausibility of even the most basic components of NMD. Second, opponents warned that if the United States proceeded with development and field testing of NMD systems, it would violate the ABM Treaty and undermine the global norm of nonproliferation. Even former president Jimmy Carter spoke publicly against Bush administration plans to abrogate the treaty.

Some of the most vocal opponents of the U.S. program were foreign governments, who grew concerned about the militarization of space and the instability NMD might cause in the international order. European allies expressed skepticism about Bush administration plans in 2001 and sought reassurance from the new president regarding the U.S. commitment to cooperative security arrangements. Meanwhile, the Russian government characterized the ABM Treaty as the "cornerstone of nuclear arms control." President Vladimir Putin publicly warned that if the United States unilaterally abandoned the treaty, Russia would consider three decades of arms control treaties null and void and would build a new generation of multiple-warhead missiles. China took a similar stand on the matter. On July 16, 2001, President Putin and Jiang Zemin, the president of China, signed a friendship and cooperation treaty that included joint statements of opposition to U.S. defense policy initiatives including missile defense.

In the wake of the events of September 11, 2001, however, the dynamics of this controversy appeared to change. President Bush has signaled his intention to move forward with NMD by proposing a four-year, $8.3 billion research budget for the system. On December 13, 2001, Bush announced that the United States would unilaterally withdraw from the 1972 ABM Treaty within six months. Although President Putin called the decision to abandon the treaty a mistake, he did pledge to maintain relations and to negotiate further reductions of nuclear arsenals. On May 24, 2002, Bush and Putin signed a treaty to reduce the number of strategic nuclear warheads held by each country to between 1,700 and 2,200 in ten years, and they hailed this achievement as the beginning of a new strategic partnership. On a parallel track, the

Pentagon began construction of ground-based facilities in Alaska for a missile defense shield in the spring of 2002 (including a command center, radar stations, and silos for the new interceptor rockets). The U.S. military also conducted another successful test of an interceptor designed to shoot down incoming ballistic missiles and planned more tests of elements of the original SDI program that might be incorporated into the new NMD. Plans for testing these systems run from 2002 to 2012.

■ CONCLUSION: PROSPECTS FOR THE FUTURE?

The proliferation of weapons is truly a major challenge to global security, but significant initiatives have been undertaken to respond to the challenge. One of the most important catalysts of global proliferation was the Cold War arms race between the superpowers. As a new decade unfolds, many scholars and politicians are taking a new look at incentives for proliferation in the post–Cold War era, and optimists say that we may be headed toward a nuclear-free twenty-first century. They argue that a global build-down in tensions—a reverse proliferation—has occurred with the end of the Cold War standoff. They cite the completion of the first Strategic Arms Reduction Treaty (START I) between the United States and the Soviet Union in January 1991, the START II agreement to reduce nuclear arsenals to 3,500 warheads each, and surprising recent agreements between Bush and Putin as evidence of progress toward a minimal nuclear deterrent relationship.

Pessimists warn, however, that many arms control initiatives are doomed to failure in the realities of a proliferating world. New and complex debates have emerged about whether defense and deterrence represent a more effective, pragmatic response to the spread of weapons technology. Strategists suggest that preparedness for a rogue state nuclear launch at the United States or a concentrated toxic terrorist attack would be the best use of government resources. The September 11 attacks and subsequent international developments appear to have bolstered this pessimistic vision, and the current war on terrorism seems to be one definitive answer to the question of how to respond to proliferation threats. Furthermore, many key officials in the Bush administration view international nonproliferation agreements with a jaded eye. U.S. government opposition to the CTBT and verification regime initiatives for the Biological Weapons Convention, combined with support for ballistic missile defense, all suggest a new attitude toward proliferation.

Regardless of one's stand in this debate, it is clear that new solutions to the proliferation challenge might be found in the twenty-first century. Citizens of the world could agree to build on the momentum of past progress by making moral and principled stands against proliferation. For instance,

Oscar Arias, the 1987 recipient of the Nobel Peace Prize, has called for a global agreement to stop arms sales to countries that have violated human rights. Others call for strengthening the small arms trade control agreement and ratification of the CTBT. These certainly would be important steps in a global effort to address ethical and moral concerns about weapons proliferation. Furthermore, moral stands against proliferation in favor of economic development, health care, and education may pave the way toward increasing peace and justice for many regions of the world in the twenty-first century.

■ DISCUSSION QUESTIONS

1. In your opinion, which of the four types of proliferation represents the most serious threat to international security?
2. Is the proliferation of conventional weapons a challenge that the global community can ever fully meet? Why or why not?
3. Is it possible that weapons proliferation could actually make the international system more stable in the twenty-first century? How might this occur?
4. What are some of the efforts that individual countries and international organizations have made to respond to the proliferation challenge? Which are most effective, and why?
5. What are some of the implications of the trade-off between expenditures on defense and social welfare programs? Can countries afford to enjoy a "peace dividend" in the post–Cold War era by diverting large sums from defense expenditures to other needs? Can they afford not to?
6. What can governments do to confront threats of terrorism using weapons of mass destruction? How have recent international developments changed the definition of the Type IV proliferation threat? Why?
7. Should the United States develop a ballistic missile defense system? What are the scientific debates regarding the feasibility of such a system?
8. In your opinion, should government leaders offer to pursue complete WMD disarmament? Why or why not?

■ SUGGESTED READINGS

Bailey, Kathleen C. (1993) *Strengthening Nuclear Non-Proliferation.* Boulder, CO: Westview Press.

Jones, Rodney W., Mark G. McDonough, Toby F. Dalton, and Gregory D. Koblentz (1998) *Tracking Nuclear Proliferation: A Guide in Maps and Charts.* Washington, DC: Carnegie Endowment for International Peace.

Karp, Aaron (1994) "The Arms Trade Revolution: The Major Impact of Small Arms," *Washington Quarterly* 17 (autumn).

Mendelsohn, Jack (2001) "Is Arms Control Dead?" *Issues in Science and Technology* 17, no. 3 (spring).

Michalak, Stanley (2001) *A Primer in Power Politics.* Wilmington, DE: Scholarly Resources, Inc.

Moodie, Michael (1995) "Beyond Proliferation: The Challenge of Technology Diffusion," *Washington Quarterly* 18 (spring).

Perkovich, George (1999) *India's Nuclear Bomb: The Impact on Global Proliferation.* Berkeley: University of California Press.

Sagan, Scott D., and Kenneth N. Waltz (1995) *The Spread of Nuclear Weapons: A Debate.* New York: W. W. Norton.

Tucker, Jonathan B. (2000) *Toxic Terror: Assessing Terrorist Use of Chemical and Biological Weapons.* Cambridge, MA: MIT Press.

3

NATIONALISM

John K. Cox

Nationalism is a complicated and widespread phenomenon in modern politics and cultures. At its most basic level, it is a sense of identity felt by individuals and groups. This sense of belonging links the individual to a group of people on the basis of certain shared characteristics. Most important among these are a common language, a common history, and common customs or cultural traditions (sometimes including religion). When this sense of identity becomes a political force, as it usually does, it generally justifies independence for the national group. This quest for political independence, often called "self-determination," is based on the perceived right of every nation, or people, to rule itself. Theoretically, this means that all the various countries, or states, of the world would become "nation-states" (independent countries composed of members of a single national group), once their populations have nationalist feelings. Making the borders of countries and nations congruent, however, is a very complicated procedure, since most empires and countries have historically contained many different ethnic or national groups and have been based on dynasties or religion or conquest or other factors instead of nationalism. Thus, only a small fraction of the world's countries today are *true* nation-states, although this process has moved further in Europe, where it began, than in other parts of the globe.

■ TYPES OF NATIONALISM

It is helpful to categorize types of nationalism according to defining concepts: What is a *nation?* Who belongs to it? And who is an outsider? In general, nationalism is broken down into two types.

■ Civic Nationalism

The first, and oldest, type was initially associated with Western European or North American politics and with countries elsewhere that followed them. It is usually called "civic," or political, nationalism, and it is seen above all as a "legal-political concept," or as a "political configuration" (Bojtar 1988: 254). Although there are competing theories regarding the origin of civic nationalism, it is Napoleon Bonaparte, who ruled France after the French Revolution (from 1799 to 1815) who is usually credited with introducing this modern concept of nationalism. As the great French novelist Andre Malraux pointed out, Napoleon convinced the French that France existed as a coherent whole, that is, he brought the diverse people of this geographic region to identify with one another and the state. Playing on this common identity became the source of much of his popularity and power; and tapping into this modern identity became a political technique (as well as a cultural achievement) that was then much copied in the rest of Europe, at least after the massive wars unleashed by Napoleon had subsided. This new national feeling went far beyond simple patriotism, which is the love of one's homeland or home region. Patriotism has been a part of human behavior since the beginning of history. But it was a rather narrow idea compared to the more inclusive civic, or modern, nationalism.

These revolutions are thus important milestones on the path to democracy, since they resulted in breaking the stranglehold on political power of the kings and aristocrats who had governed up to this time. Still, the much-heralded civic nationalism can also be exclusive. For instance, the U.S. Constitution was designed in the 1780s to deny women and slaves the right to vote. It was only after the Civil War that African American men were officially given the right to vote (the Fifteeth Amendment), and in many states this right was not protected by meaningful enforcement of laws until the Civil Rights Acts of the 1960s, almost 100 years later. Women were denied the right to vote almost everywhere until the twentieth century; in the United States, this right was provided by the Nineteenth Amendment in 1920. In theory, though, civic nationalism assumes that citizenship and nationality are identical (Liebich, Warner, and Dragovic 1995: 186). The nation is a political population, united in its ideas and habits.

Most scholars who deal with nationalism—historians, political scientists, sociologists, and, increasingly, psychologists—believe that the growth of nationalism is a fundamental aspect of modernization. Generally modernization involves industrialization, urbanization, increased literacy, and secularization. This was as true of European history in the nineteenth century as it was of the decolonizing world—mostly Africa and Asia—in the twentieth. Therefore, the growth of nationalism involves two processes: its appearance in people's minds as a sense of identity, belonging, and loyalty; and its growth (or cultivation) into a political force, which ultimately works to create a sovereign state. One of the most modern aspects of nationalism is that differences in socioeconomic status (class) are minimized; often religion is de-emphasized, too, and sometimes so is race. This creates a bigger and more powerful body of subjects or citizens; in positive manifestations of nationalism, this "people," or nation, will embrace cultural diversity and try to rule itself by increasing civil liberties and democratic voting procedures.

■ Ethnic Nationalism

The other type of nationalism, "ethnic" nationalism, was originally associated with countries in Eastern and Central Europe. This nationalism is based on "ancestral association" (Bojtar 1988: 254) as compared to civic nationalism, which can embrace diverse people who live within shared borders. Ethnic nationalism requires a common culture, way of life, and above all a perceived sense of genetic links (as in a greatly extended family) among the members of the ethnic community. The word *ethnic* comes from the Greek word *ethnos* meaning a group of people united by their common birth or descent. It should be noted that all types of nationalism are in some way exclusionary. If nothing else, this is true because of the presence of borders and frontiers. But ethnic nationalism, due to its emphasis on the "blood line" or racial connections among citizens, is far more exclusionary than civic nationalism and pays less attention to political boundaries.

The historical differences between these types of nationalism are great and remain relevant to this day. The war in Bosnia-Herzegovina can be better understood by remembering that many Serbs and Croats adhere to the exclusive nationalism of the second category (see the case study on Yugoslavia on p. 44). Why? Because the more inclusive civic nationalism of Western Europe developed in the spirit of certain key turning points in European civilization, such as the Enlightenment and the growth of middle-class democracy. Western European nationalism arose in societies that were already modernizing, while the peoples of Eastern Europe were neither independent nor economically modern. In short, Eastern Europe became nationally conscious before it had experienced economic development, representative government, and political unity (or in many cases even independence from foreign rule). A central result was the desire to alter the political boundaries to coincide with national or cultural boundaries (Sugar and Lederer 1994: 10; Kohn 1965: 29–30); another result was to embrace a greater sense of exclusivity in determining who was "in" and who was "out" of the nation.

■ THE EVOLUTION OF NATIONALISM
SINCE THE NINETEENTH CENTURY

Despite the differences between civic and ethnic nationalism, both concepts have much in common. As pointed out at the beginning of this chapter, both lead to a sense of belonging and eventually the desire for political independence. Both have tended to foster "popular sovereignty," or democracy; and both have been used by leaders to galvanize support for the state (as opposed to a monarchy, in which the people were merely "subjects" instead of "citizens").

During the nineteenth century, more and more European political leaders were starting to make use of the great political power of nationalism. Nationalism appealed to many with its symbolism of national unity and mission.

Skeptics and detractors of the movement, however, had strong arguments against it. Leaders of Europe's many multinational states (the United Kingdom, Russia, the Ottoman and Habsburg Empires), the nobility, many leaders of the Roman Catholic church, and Marxists, all for different reasons, opposed nationalism.

Despite this diverse and often intense opposition, the twentieth century began with what most people regard as the triumph—or, as some would say, the running amok—of the national idea. The great powers of Europe, such as Britain, Germany, Russia, and France, became imperialistic and sought to ratchet up their power considerably, both in Europe and abroad. The result was World War I (1914–1918), in which massive armies and new military technology combined with the propaganda of national glory and the vilification of the enemy to create a new level of battlefield fury and destructiveness. The Great War, as it is also called, was the first total war involving horrific new violence using poison gas, tank assaults, long-range and high-powered artillery, submarine warfare, the use of flamethrowers and machine guns, and the bombardment of civilian population centers.

There is another important connection between the Great War and nationalism: The number of nation-states in Europe was greatly expanded as a result of the peace treaties at its conclusion. The old multinational empires of Europe collapsed (except for the United Kingdom). In their place arose a set of what diplomats endorsed at the time as nation-states, from Finland in the north to Turkey in the south. These included Poland, Estonia, Latvia, Lithuania, Hungary, and Austria as well as two small confederations of related peoples—Czechoslovakia and Yugoslavia.

The study of nationalism is growing increasingly sophisticated. Some scholars are now examining the role of territory and geography in creating psychological identity and loyalty to the state, while others view nationalism as a kind of expanded male-dominated hierarchy (also known as patriarchy), which extends men's control over women, especially in the areas of work and sexuality. A recent addition to the analytical concepts used to understand nationalism is "nested identities," a framework for studying the way individual people can maintain and prioritize various kinds of identities relating to various parts of their lives.

■ FUNCTIONS OF NATIONALISM

Nationalism functions in five ways. First, there is the matter of *identification,* whereby individuals consider themselves, especially since the advent of industrialization and its processes of urbanization and secularization, to be part of a nontraditional mass group, the "nation." Second, governments since the time of Napoleon have used nationalism as a means to *mobilize* military and economic power and to further their own legitimacy. Third, nationalism can function as a *centrifugal force* when it breaks up bigger countries (or empires) into smaller ones. This occurred in many European countries after World War I. It also took place in a massive way in the British Empire after World War II when India, Pakistan, Ghana, Nigeria, and other former colonies became independent. Then in the 1990s, it occurred again in the breakups of Czechoslovakia (into the separate Czech and Slovak republics) and Yugoslavia (into Slovenia, Croatia, Bosnia, Macedonia, and Serbia-Montenegro). Canada and Spain are two countries experiencing the centrifugal effects of separatist nationalism today: the French-speaking Quebecois in Canada and the Catalonians and Basques in Spain. The recent civil war in the Democratic Republic of Congo (formerly Zaire) is another important example of different national groups competing for power and gradually crippling the power of the central government.

Nationalism can also work in a fourth way, as a *centripetal force*, when it unites various people into new nation-states, such as occurred in the long and bloody unification struggles of the Germans and Italians in the nineteenth century or in the Vietnam War in the twentieth century.

Fifth, nationalism can serve as a form of *resistance,* especially to colonial intruders. In Africa, the Middle East, and Asia, this has often been a kind of state-run, top-down nationalism that aims at organizing more meaningful resistance to actual or potential invaders. Sometimes this top-down nationalism is called "reform nationalism" (Breuilly 1993: 9). In Turkey after World War I, Mustafa Kemal Ataturk launched a highly successful plan of economic and political modernization based on this kind of government-led reform nationalism. Cuba under Fidel Castro fits this definition as well. Another kind of resistance to colonialism takes the form of wars of independence (sometimes called national liberation struggles). Important examples of this kind of national struggle include the Vietnam War and the

Algerian war of independence (1954–1962). *Peripheral nationalism* is a related phrase used to describe the impetus given to national groups that are emerging from a centrally run empire that is collapsing; geopolitical concerns are important here, since local leaders must step into a power vacuum and create structures of governance that are both effective and recognized as legitimate by the population.

■ NEGATIVE ASPECTS OF NATIONALISM

As discussed earlier, nationalism can be an individual's sense of identity, a political allegiance, and a force for military and political change. Arising from these different levels of meaning are various negative effects of nationalism. Many of the conflicts in the world today originate in national disputes. A quick glance over the headlines shows warfare, ethnic conflicts, or genocide in Bosnia, Chechnya, Rwanda, Indonesia, Canada, South Africa, Macedonia, Cyprus, Israel, and Ireland. One can discuss these negative, conflict-producing effects of nationalism in terms of the following categories: imperialism, the glorification of the state, the creation of enemies, the overlap with religion, discrimination against minorities, and competing rights.

■ Imperialism

Self-confidence and group assertiveness, integral aspects of nationalism, can lead to arrogance or aggressiveness. Imperialism, which is the projection of a country's power beyond its borders to achieve the subjugation or exploitation of another country, is as old as history itself. But it takes on greater intensity when it meets with a sense of national unity and purpose. The "scramble for Africa" of the late nineteenth century, when many European states collaborated in literally carving up and occupying most of the continent, is a breathtaking example of arrogant imperialism imbued with a purported civilizing mission, or "white man's burden," which justified the exploitation of other races. Carried to a much greater extreme, nationalism can end in genocide, as it did in the wildly homicidal policies of Adolf Hitler in the Third Reich, who sought to rid the world of Jews in order to make it "safe" for Germans.

■ Glorification of the State

Although many early nationalists, especially in the nineteenth century, believed that the nation-state was a vehicle of progress and liberty for all human beings, not all nationalist thought is connected with individual freedom. Indeed, nationalism often encourages antidemocratic practices. When a "people," or nation, feels threatened by neighbors, or when it has a history of underdevelopment or division, political leaders can make the case for an authoritarian (antidemocratic) government. Sometimes, in the case of fascist governments, which are extremely authoritarian and stress anti-individualism, racial or national homogeneity, scapegoating, and militarism, the state or its leader comes to be regarded as the ultimate expression of the people's character and ambitions (Payne 1995; Weber 1964). Loyalty to governments like these is extremely dangerous; democracy is shortchanged at home and foreign policy is often aggressive.

■ Creation of Enemies

Another negative effect of nationalism can take place at the most basic level of self-identification. When people identify with one group, they often develop mistrustful or hostile feelings about people outside that group. Even neighboring states with a great deal in common can come to mistrust each other, as in the case of the recent fishing controversies between the United States and Canada. Similarly, countries with common political interests and similar economic systems—such as the United States and Japan—can develop deep misunderstandings based largely on national feeling.

■ Overlap with Religion

In some conflicts, such as those in Northern Ireland, the Middle East, and Bosnia, nationalism and religion cross paths in a destructive way. In the current three-way struggle in Bosnia, between mostly Orthodox Christian Serbs, Roman Catholic Croats, and Bosnian Muslims, religion factors heavily. Adding a religious dimension to nationalism can intensify divisive feelings; for instance, it can sanction killing—or dying—for a cause. Thus, it can make nationalists more fanatical and conflicts bloodier (Landres 1996).

■ Discrimination Against Minorities

Other difficulties arise when states or countries are actually constructed on national principles. Such principles hold that only members of a given national or ethnic group have the right to live in the new national state. Often a related principle tends to hold a lot of weight also—for example, the belief that only members of a particular ethnic group should enjoy the full benefits of citizenship. This creates a problem for minority groups. Major examples include the Hungarians in Romania and Slovakia; the Russians in the former Soviet republics (now independent states) of Estonia, Latvia, and Lithuania; and, until the creation of the Irish Republic, the Irish in the United Kingdom.

■ *Competing Rights*

Another negative aspect of nationalism lies in the competing rights and claims that states make against one another. Three kinds are derived from or have a major impact on ethnic and minority questions. The first involves *historic rights*. These include claims by one national group to a certain piece of territory based on historical precedent. In other words, who was there first? This issue is hotly debated in Transylvania, a large portion of western Romania that has a substantial Hungarian population. In Bosnia, the competing parties of Serbs, Bosnian Muslims, and Croats have each tried to prove that they contributed more to the region's cultural heritage and, having set the tone for the region's culture, deserve to mold the region's political future now.

Next are *ethnic rights,* which address the question of who is currently in the majority in a given region. The contemporary setting—determined by population counts, polls, and votes of self-determination (such as in the Austrian province of Carinthia and in the Polish-German region of Silesia just after World War I)—is the decisive factor, not the complicated historical record of settlements, assimilation, immigration, and emigration.

The final claim can be referred to as *strategic rights*. Sometimes a state will claim a piece of territory simply because it needs that territory in order to be viable. This usually means the land is necessary for the country's defense or basic economic well-being. For example, after World War I the victorious Allies gave the new state of Czechoslovakia the Sudetenland region, even though it was heavily populated with Germans. This was done to provide the fledgling republic with a more mountainous, defensible border. Unfortunately, the Nazi dictator Adolf Hitler would later attack Czechoslovakia both diplomatically and militarily to "liberate" the Germans of that region, who, he claimed, were being denied their right to self-determination.

■ CHALLENGES TO NATIONALISM

Historically nationalism has been opposed by many forces. In addition to the conservative opposition in Europe, European imperialist powers in Africa—especially the United Kingdom, France, Belgium, and Portugal—resisted the growth of nationalism in their colonies. They did this despite being more or less nation-states themselves. This is because nationalism among colonized peoples presented a direct challenge to European domination and exploitation.

I will now examine the four main challenges to the nation-state. One of these challenges is inherent in the ideal of nationalism itself. This is the problem of carrying the principle of self-determination through to its logical conclusions; if one national group deserves its own country and independence, then do not all groups deserve these things too? But countries are destabilized when every ethnic group within them agitates for its independence. And sometimes so-called microstates are created that are too small to be economically viable and that swell the membership of the United Nations and affect voting patterns there. For instance, the Pacific island country of Kiribati has about one-eighth as many people as the Canadian city of Toronto; likewise, the combined populations of thirty-eight microstates total only about a third of that of California (Rourke 1995: 201–202).

A related issue is *devolution*, or the decentralization of power in ethnically mixed countries. This usually does not result in the breakup of the country. The United Kingdom continues to experiment with this principle by giving more and more autonomy to its Welsh and Scottish regions. Belgium has also achieved a balance, based on this principle, between its Flemish and Walloon populations. Russia is faced with this issue early in the twenty-first century in many autonomous regions and districts.

Second is the issue of *supranational* groupings of various kinds. At the height of the era of decolonization, some Arab and African countries tried to establish political leagues that cooperated on a wide variety of issues. In 2002 there are regional political and economic groupings on every continent. But in Europe, the blossoming of the European Union seems to herald an age of ever greater integration of nations. The United Nations, of course, while generally respecting the sovereignty of all countries, is the best example of a global group above the national level. Other contemporary examples include the North American Free Trade Agreement (NAFTA) and the Common Market of the South (Mercosur); both are regional free trade groups in the Western Hemisphere.

Third, modern *economic developments* are also undermining the nation-state. The influence of multinational corporations, the obsession with free trade, and the appearance of a global, computer-driven, mass market economy are breaking down barriers among populations and eroding the sovereignty of smaller, less developed nations. This trend is analyzed in detail in Benjamin Barber's *Jihad vs. McWorld* (1996), where a grim picture is painted of an increasingly standardized, shallow world culture dominated by a few, nearly all-powerful, marketing agents and producers of consumer goods.

In the 2000s the concept of national identity, and even to some degree the concept of nation-states, is in flux. The prevalence of computer-driven communication on the Internet and the World Wide Web affects society in many ways. From shopping to political discussions to dating networks, geography and distance are suddenly rendered virtually inconsequential by computers. The much-heralded "global village" of travel and communication has to some extent arrived, although its effects will likely never be as graciously positive and progressive as the gurus of technology predicted a few decades ago. Computer culture has developed rapidly along with the general economic shift in the world's most developed countries (such as Germany, the United Kingdom, the United States, Canada, and Japan) into service-based economies (in contrast to economies based on the production of industrial goods). Many important changes in thought and attitudes go along with these technological and economic shifts. Service economies are oriented toward individual consumption, and the Internet means that individuals can have a maximum of self-fulfillment with a minimum of real contact with other citizens. This can reduce the sense of group loyalty so important in nationalism.

Finally, the world is also witnessing a revival of conservative religious activity. This is most prominent in the Muslim world, but it is also present in other religions. Politically speaking, it is the new Islamism (sometimes called "Muslim fundamentalism") that most affects international politics, because it rejects capitalism and the decadence of Western culture as manifestations of a new imperialism. Since much, although by no means all, of the Muslim world consists of states that are ethnically and linguistically Arab, there is added potential for cooperation that transcends political boundaries. The supranational nature of the terrorism of Islamist extremists was evident in the deadly attacks on the United States in September 2001, in which terrorists from Saudi Arabia, Egypt, and other countries joined together—probably with the support of other governments such as the Taliban in Afghanistan—to strike a blow for an agenda involving issues in Iraq, Israel, Saudi Arabia, and many other countries both inside and outside the Middle East.

The following two case studies illustrate current national conflicts in various parts of the world. They give us an idea of how national issues mix with other kinds of problems to create major crises.

■ CASE STUDY ONE: YUGOSLAVIA

The region of southeastern Europe known as the Balkans provides numerous intriguing case studies of nationalism at work. One of the characteristics of the region is the prevalence of ethnic, or cultural, nationalism rather than civic, or political, nationalism. Another is the highly diversified nature of its population. In many areas of the Balkans numerous ethnic or national groups live closely together; groups often intermingle and sometimes occupy the ancestral homelands of their neighbors. Two of the most mixed of these areas are Bosnia and Macedonia, both of which were part of the former Yugoslavia.

A third major characteristic of Balkan societies is a long history of foreign rule. Various empires, from the Ottoman and Habsburg to the Russian and Soviet, have dominated the region, preventing the self-determination of Balkan peoples. The two main peoples within the multinational state of Yugoslavia were the Serbs and Croats. It was their conflicting national aspirations—strengthened and made poisonous, many would say, by their leaders—that provided the impetus for the breakup of the country in 1991–1992.

The term *Yugoslavia* means simply "land of the South Slavs." The country was created in 1918 as a kind of catch-all state for a number of small nationalities, including several that had been part of empires that collapsed in World War I. Thus, the term *Yugoslav* did not correspond to any genuine national or ethnic group; it was a matter of citizenship only (civic nationalism), except for a small number of idealists or people who were part of mixed families created by marriages between members of different national groups. The number of Yugoslav citizens who identified with Yugoslav nationality (as opposed to the older categories of Serb, Croat, etc.) never surpassed 10 percent, although it was highest in Bosnia and Vojvodina (a province of northern Serbia), where old communities of varying nationalities lived in the closest proximity to each other.

During its existence, the country—first under the authoritarian rule of the Serbian royal family and then after World War II under the firm hand of the Communist military leader Josip Broz (known as Tito)—was divided into provinces or "republics" that reflected its chief national groups: Serbs, Croats, Bosnian Muslims, Slovenes, and others. There were also large and important minority groups, especially Albanians in the southern regions.

Rivalries among the various South Slavic national groups have been common, as they are among almost all neighboring peoples. But the frequently used journalistic phrases "ancient ethnic hatreds" and "long-smoldering ethnic feuds" are not accurate. While the Muslim-Christian rivalry in the Balkans had been a problem since the Middle Ages in Bosnia, the animosities between Serbs and Croats became acute only during World War II.

After the Nazis and their allies carved up Yugoslavia in 1941, puppet states in both Croatia and Serbia emerged (see Figure 3.1). Both countries, but especially Croatia, sought to expand their territory and to homogenize their population at the expense of their neighbors and minorities. Further complicating the situation was the nature of Yugoslav resistance to the Nazis, which was led by the Communists under Tito but which included other rival political groups.

Tito's post–World War II government sought to stabilize the country's national groups by one-party rule and by a decentralized administration. However, in the early 1990s Yugoslavia began to break apart when Slovenia, Croatia, and Bosnia-Herzegovina sought and won their independence. Thus, the effort to create a single country (based on civic nationalism) among disparate ethnic groups failed. Unrest in Bosnia continues, and in

Figure 3.1 Yugoslavia and Its Successor States

Source: Reprinted from Wayne C. McWilliams and Harry Piotrowski, *The World Since 1945: A History of International Relations,* 5th ed. (Boulder, CO: Lynne Rienner, 2001). © Copyright 2001 Lynne Rienner Publishers.
    Note: Serbia, Montenegro, and Kosovo constitute what remains of Yugoslavia as of 2002.

1999 the United States led its NATO allies into war with Serbia over Slobodan Milosevic's "ethnic cleansing" of the Albanians in the province of Kosovo. In some ways, Tito may have made the national situation worse. Still, it is impossible to attribute the breakup of the country to any one cause. Nationalism played a part, as did economic problems, the ambitions of current leaders, and the failure of the Communists to allow or promote a pluralistic civil society that could have taught deeper loyalties to the central government and the Yugoslav ideal.

    Doubtless the map and distribution of population affected the way the country fell apart, too. The Serbs were always the dominant group, both numerically and politically, in Yugoslavia. But at 36 percent of the population, they did not form a majority in the 1980s. Other groups, especially the Croats, were numerically strong enough to challenge them; the smaller Slovene population wielded considerable economic power, and Tito favored other small groups such as the Bosnian Muslims and Macedonians, whom he viewed as brakes on potential Serbian dominance. Furthermore, of the country's 8 million Serbs, almost 25 percent of them lived outside the borders of Serbia, mostly in Bosnia and Croatia. This meant that when those two countries began to move away from the rest of Yugoslavia in 1991, nationalists in Serbia were aroused to war to try to keep the country together. As imperfect as Serbs thought Tito's Yugoslavia had been, at least all their people had lived in one country for the first time since the fourteenth century and they did not need to fear a repeat of Croatian violence as seen in the 1940s.

The ethnic drama continues to unfold in the former Yugoslavia as troops from the North Atlantic Treaty Organization (NATO) continue to occupy both Kosovo and Macedonia in an effort to stop the killing. Kosovo, the Albanian-dominated southern province of Serbia, became an international protectorate as did Macedonia, referred to as the Former Yugoslav Province of Macedonia. The latter is threatened by its minority Albanian population and their fellow Albanians in Kosovo. At the same time that these efforts are taking place, Slobodan Milosevic, the recently defeated president of Yugoslavia, has been extradited to The Hague, Netherlands, to stand trial for genocide, crimes against humanity, and war crimes. See Chapter 4 on human rights for more about these charges.

■ CASE STUDY TWO: THE ARAB-ISRAELI CONFLICT

The contemporary troubles between Arabs and Israelis form the main, but not only, crisis in the Middle East. This conflict centers on possession of the territory known as Palestine, which is important in both a religious and a historical sense to both Arabs and Jews. Since the diaspora (forcible dispersion) of Roman times, most Jews have lived outside this ancestral homeland. Arabs were the majority, though not the only, population group in Palestine during the intervening centuries. During World War I, the British, who were also cooperating with the Arabs, issued the Balfour Declaration in support of a Jewish state there. In 1948, when Palestine was partitioned and the state of Israel was created (see Figure 3.2), hundreds of thousands of Arabs were displaced. As a result, many Muslim states have, at least until recently, refused to recognize Israel's right to exist.

Wars between Israel and its Arab neighbors broke out in 1948, 1956, 1967, and 1973. Although Israel won all the wars, the situation remained volatile in large part because of the Cold War. The Soviet Union supported Arab states, while the United States aided the Israelis. In 1964, the Palestine Liberation Organization (PLO) was formed. Led by Yasir Arafat, the PLO has operated as a refugee organization, a government-in-exile, and a

Figure 3.2 The Expansion of Israel

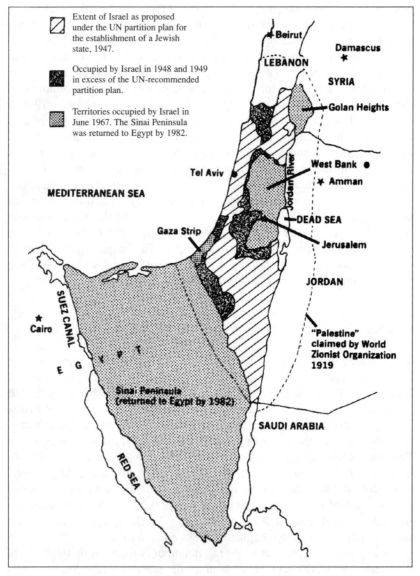

Source: Reprinted from Wayne C. McWilliams and Harry Piotrowski, *The World Since 1945: A History of International Relations,* 5th ed. (Boulder, CO: Lynne Rienner, 2001). © Copyright 2001 Lynne Rienner Publishers.

terrorist group; today it has been granted partial state power in several regions in and around Israel.

In addition to the territorial, religious, and linguistic differences between the Arabs and the Israelis, other factors have combined to make the nationalism of the conflicting parties more intense. Some Jews began to take an active interest in returning to their traditional homeland in the 1890s, due to increasing anti-Semitism in Europe. The movement known as Zionism proclaimed the Jewish people a nation with territorial claims to Palestine and not just a religion. In the 1940s, the terrors of the Holocaust gave added impetus to Jewish emigration.

On the Arab side, the issues involve more than just the fate of the Palestinian Arabs; many Arabs view Israel as an outpost of Western imperialism. Long dominated by foreign powers, the Arabs have revealed "frustration over past and present weakness" (von Laue 1987: 350). Since World War II, the Arab countries have cooperated in various international bodies designed to foster unity and common purpose. These include the Arab League (since 1945), the Organization of Arab Petroleum Exporting Countries, or OAPEC (since 1960), and the short-lived United Arab Republic, a fusion of Egypt and Syria.

In the mid-1990s, the peace process in the Middle East began making significant progress. The Israeli government was finally recognized by the PLO, Egypt, Jordan, and other key Muslim states. And with this recognition of Israel's right to exist, the Israeli governments of Yitzhak Rabin (assassinated in 1995) and, later, Ehud Barak were able to push ahead with a "land-for-peace" deal with Yasir Arafat of the PLO.

Yet tragedy struck in 2000. As Israeli withdrawals from disputed territory in the West Bank were starting and Arafat's Palestinian Authority was assuming more autonomous governmental functions, a second intifada, or uprising (the first was in 1987), broke out. The violence—suicide bombings and shootings carried out by radicals from the unofficial Palestinian groups Islamic Jihad and Hamas—escalated as the Israeli military responded with increasing force. By 2002, the level of bloodshed was continuing to mount, and many outside observers doubted whether Arafat retained the ability to control radical factions among his people. Arafat supporters say that the Israelis and their American allies have demanded too much of Arafat without giving him concessions to raise the Palestinians' confidence in Israel's good intentions. The peace process at the time of this writing might, sadly, be effectively sabotaged.

Like the Yugoslav case, nationalism has mixed with other historical and political issues such as population and Cold War rivalry. Given the complexity of this situation, it is no wonder that the main actors, even with the assistance of the international community, have not been able to solve the Palestinian issue.

■ CONCLUSION

As mentioned, nationalism has been one of the most powerful forces in the world over the past 200 years. It operates in various ways on individuals, social groups, and governments. Nationalism can build states or destroy them; it can provide individuals or organizations with identity, motivation, and justification for their actions. Although it is true that nationalism provides important links between people and can give them a common purpose (as in patriotism), nationalism also causes conflicts and can be used to justify aggression and feelings of superiority. Thus, it has a very mixed reputation in terms of its positive and negative effects. While most people regard nationalism as an inevitable companion to social modernization, there are some signs that the usefulness or relevance of nationalism may be on the wane.

■ DISCUSSION QUESTIONS

1. Which of the functions of nationalism do you think is the most important, and why?
2. What are some of the negative aspects of nationalism? Which do you think is the most dangerous or immoral?
3. Why do some countries have civic nationalism and others ethnic nationalism?
4. Which of the two types of nationalism is most prevalent in the United States?
5. Do you think nationalism will be replaced by other allegiances in the next few decades?

■ SUGGESTED READINGS

Breuilly, John (1993) *Nationalism and the State*. Chicago: University of Chicago Press.

Brown, Michael E., Owen R. Coté, Sean M. Lynn-Jones, and Steven E. Miller, eds. (1997) *Nationalism and Ethnic Conflict: An International Security Reader*. Cambridge, MA: MIT Press.

Eley, Geoff, and Ronald Grigor Suny, eds. (1996) *Becoming National: A Reader*. New York: Oxford University Press.

Geenfeld, Liah (1992) *Nationalism: Five Roads to Modernity*. Cambridge, MA: Harvard University Press.

Gellner, Ernest (1983) *Nationalism*. Ithaca, NY: Cornell University Press.

Hechter, Michael (2000) *Containing Nationalism*. New York: Oxford University Press.

Hobsbawm, Eric (1992) *Nations and Nationalism Since 1780: Programme, Myth, Reality*. New York: Cambridge University Press.

Hutchinson, John, and Anthony D. Smith, eds. (1996) *Ethnicity.* New York: Oxford University Press.

———, eds. (2000) *Nationalism: Critical Concepts in Political Science.* New York: Routledge.

Kecmanovic, Dusan (1996) *The Mass Psychology of Ethnonationalism.* New York: Plenum.

Liebich, André, Daniel Warner, and Jasna Dragovic, eds. (1995) *Citizenship East and West.* London: Kegan Paul International.

Sugar, Peter F., and Ivo John Lederer, eds. (1994) *Nationalism in Eastern Europe.* Seattle: University of Washington Press.

Teich, Mikulaš, and Roy Porter, eds. (1993) *The National Question in Europe in Historical Context.* Cambridge: Cambridge University Press.

Tilly, Charles, ed. (1975) *The Formation of National States in Western Europe.* Princeton, NJ: Princeton University Press.

4

THE CHANGING FACE OF GLOBAL HUMAN RIGHTS

D. Neil Snarr

On December 10, 1948, the General Assembly of the United Nations (UN) approved the Universal Declaration of Human Rights (UDHR). There were no votes against the document, but eight countries abstained while forty-eight voted in its favor. Since that time the UN has approved well over 200 documents that elaborate and expand these rights (such agreements are often referred to as conventions, treaties, or covenants). Since their inception human rights issues have received tremendous attention. As one member of the U.S. House of Representatives and student of human rights has put it, "The defense of internationally recognized human rights has become the most universally accepted moral standard in the world today. Across the ideological spectrum, from the far left to the far right, there is agreement that the one unifying spiritual ideal in modern society is the enhancement and enforcement of human rights" (Drinan 1987: vii). This chapter looks at these rights, the controversies that surround them, the efforts to support them, the many forces that inhibit their realization, and some specific case studies.

■ THE ORIGIN OF HUMAN RIGHTS

Declarations and agreements that contain historic steps toward the 1948 UDHR are numerous. Such documents as the British Magna Carta (1215), the French Declaration of the Rights of Man and of the Citizen (1789), and the U.S. Bill of Rights (1791) are just a few of these contributions. Late in the nineteenth century slave trade was outlawed and early in the twentieth century slavery itself was outlawed. Later, humanitarian considerations in the conduct of war were agreed upon, and the treatment of workers, prisoners,

53

and women were the subject of international agreements. The events that immediately preceded the establishment of the UN and the approval of the Universal Declaration of Human Rights, however, were World War II and the genocide against Jews and other groups in Europe. Because of these indescribable horrors, the world community founded the UN in the hopes of avoiding such wars and violations of human rights in the future.

■ WHAT ARE HUMAN RIGHTS?

What are human rights? One UN document refers to them as "inalienable and inviolable rights of all members of the human family" (UN 1988: 4). According to one scholar,

> The very term *human rights* indicates both their nature and their sources: they are the rights that one has simply because one is human. They are held by all human beings, irrespective of any rights or duties one may (or may not) have as citizens, members of families, workers, or parts of any public or private organization or association. In the language of the 1948 declaration, they are universal rights. (Donnelly 1993: 19)

How do human rights come into existence? Discussions of human rights generally start in the UN General Assembly, where they are debated, given public scrutiny, and voted on. Passage of human rights conventions in the General Assembly is the easy part; it only takes a majority vote. After the General Assembly approves these conventions, they are opened for signatures by member states; and after a designated number of countries have ratified them, they are said to "come into effect." It often takes many years for this to happen.

After conventions come into effect, the agreeing countries are expected to pass laws, if they do not have such laws, that will ensure their observance and enforcement. Eventually, it is hoped, all countries will approve such human rights laws and they will become international law. I will discuss the problems of enforcing these rights later in this chapter.

■ THE UNIVERSAL DECLARATION OF HUMAN RIGHTS

What rights have been identified as human rights? One way to approach this question is to divide the UDHR of 1948 into three generations, or classes. These three generations have different origins and represent different views of human rights. Since the 1948 covenant was approved these rights have been expanded and elaborated, but they still can serve as a place to initiate a discussion. The UDHR includes thirty articles and the first, introductory, article declares that "All human beings are born free and equal in dignity and rights. They are endowed with reason and conscience and should act toward one another in a spirit of brotherhood."

Following this statement are the first generation of rights, often referred to as *civil and political* (or first-generation rights); they are contained in Articles 2 through 21 of the UDHR. They focus on the rights of the individual and emphasize the responsibility of governments to refrain from unjustly interfering in the lives of their citizens. They are as follows:

> 2. Everyone is entitled to all the rights and freedoms set forth in this Declaration, without distinction of any kind, such as race, color, sex, language, religion, political or other opinion, nation or social origin, property, birth or other status. Furthermore, no distinction shall be made on the basis of the political, jurisdictional or international status of the country or territory to which a person belongs, whether it be independent, trust, non-self-governing or under any other limitation of sovereignty.
>
> 3. Everyone has the right to life, liberty and security of person.
>
> 4. No one shall be held in slavery or servitude; slavery and the slave trade shall be prohibited in all their forms.
>
> 5. No one shall be subjected to torture or to cruel, inhuman or degrading treatment or punishment.
>
> 6. Everyone has the right to recognition everywhere as a person before the law.
>
> 7. All are equal before the law and are entitled without any discrimination to equal projection of the law. All are entitled to equal protection against any discrimination in violation of this Declaration and against any incitement to such discrimination.
>
> 8. Everyone has the right to an effective remedy by the competent national tribunals for acts violating the fundamental rights granted him by the constitution or by law.
>
> 9. No one shall be subjected to arbitrary arrest, detention, or exile.
>
> 10. Everyone is entitled in full equality to a fair and public hearing by an independent and impartial tribunal, in the determination of his rights and obligations and of any criminal charge against him.
>
> 11. (1) Everyone charged with a penal offence has the right to be presumed innocent until proved guilty according to law in a public trial at which he has had all the guarantees necessary for his defense.
>
>   (2) No one shall be held guilty of any penal offence on account of any act or omission which did not constitute a penal offence, under national or international law, at the time when it was committed. Nor shall a heavier penalty be imposed than the one that was applicable at the time the penal offence was committed.
>
> 12. No one shall be subjected to arbitrary interference with his privacy, family, home or correspondence, nor to attacks upon his honor and reputation. Everyone has the right to the protection of the law against such interference or attacks.
>
> 13. (1) Everyone has the right to freedom of movement and residence within the borders of each state.

   (2) Everyone has the right to leave any country, including his
own, and to return to his country.
14. (1) Everyone has the right to seek and to enjoy in other countries
        asylum from persecution.
    (2) This right may not be invoked in the case of prosecutions
        genuinely arising from non-political crimes or from acts con-
        trary to the purposes and principles of the United Nations.
15. (1) Everyone has the right to a nationality.
    (2) No one shall be arbitrarily deprived of his nationality nor
        denied the right to change his nationality.
16. (1) Men and women of full age, without any limitation due to
        race, nationality or religion, have the right to marry and
        to found a family. They are entitled to equal rights as to mar-
        riage, during marriage and at its dissolution.
    (2) Marriage shall be entered into only with the free and full con-
        sent of the intending spouses.
    (3) The family is the natural and fundamental group unit of so-
        ciety and is entitled to protection by society and the State.
17. (1) Everyone has the right to own property alone as well as in as-
        sociation with others.
    (2) No one shall be arbitrarily deprived of his property.
   18. Everyone has the right to freedom of thought, conscience and religion; this right includes freedom to change his religion or belief, and
freedom, either alone or in community with others and in public or private, to manifest his religion or belief in teaching, practice, worship and
observance.
   19. Everyone has the right to freedom of opinion and expression; this
right includes freedom to hold opinions without interference and to seek,
receive and impart information and ideas through any media and regardless of frontiers. . . .
21. (1) Everyone has the right to take part in the government of his
        country, directly or through freely chosen representatives.
    (2) Everyone has the right to equal access to public service in his
        country.
    (3) The will of the people shall be the basis of the authority of
        government; this will shall be expressed in periodic and gen-
        uine elections which shall be by universal and equal suffrage
        and shall be held by secret vote or by equivalent free voting
        procedures.

   Originating in seventeenth- and eighteenth-century Western ideas, these
rights found expression in the revolutions of France, Britain, and the United
States. The United States often views these as civil rights, which the U.S.
government equates with human rights. These rights were supplemented by
the International Covenant on Civil and Political Rights, which clarify and
expand them. The covenant was open for signatures in 1966, but it took
until 1976 to acquire the thirty-five signatories required to implement the
convention. Even with the strong support of U.S. president Jimmy Carter in
the late 1970s, the covenant was not ratified by the United States until

1992. It has been this first generation of human rights that has received the most attention and seen the most success (even though during the Cold War there was very little progress).

The second generation of human rights is referred to as *social and economic* rights. Contained in Articles 22 through 26 of the UDHR, they stem from the Western socialist tradition. To some degree they have developed in response to what are considered to be the excessive individualism of the first generation of rights and the impact of Western capitalism and imperialism. They focus on social equality and the responsibility of the government to its citizens. They are as follows:

22. Everyone, as a member of society, has the right to social security and is entitled to realization, through national efforts and international cooperation and in accordance with the organization and resources of each State, of the economic, social and cultural rights indispensable for his dignity and the free development of his personality.

23. (1) Everyone has the right to work, to free choice of employment, to just and favorable conditions of work and to protection against unemployment.

(2) Everyone, without any discrimination, has the right to equal pay for equal work.

(3) Everyone who works has the right to just and favorable remuneration ensuring for himself and his family an existence worthy of human dignity, and supplemented, if necessary, by other means of social protection.

(4) Everyone has the right to form and to join trade unions for the protection of his interests.

24. Everyone has the right to rest and leisure, including reasonable limitation of working hours and periodic holidays with pay.

25. (1) Everyone has the right to a standard of living adequate for the health and well-being of himself and of his family, including food, clothing, housing and medical care and necessary social services, and the right to security in the event of unemployment, sickness, disability, widowhood, old age and other lack of livelihood in circumstances beyond his control.

(2) Motherhood and childhood are entitled to special care and assistance. All children, whether born in or out of wedlock, shall enjoy the same social protection.

26. (1) Everyone has the right to education. Education shall be free, at least in the elementary and fundamental stages. Elementary education shall be compulsory. Technical and professional education shall be made generally available and higher education shall be equally accessible to all on the basis of merit.

(2) Education shall be directed to the full development of the human personality and to the strengthening of respect for human rights and fundamental freedoms. It shall promote understanding, tolerance and friendship among all nations, racial or religious groups, and shall further the activities of the United Nations for the maintenance of peace.

(3) Parents have a prior right to choose the kind of education that
shall be given to their children.

Rather than protecting the citizen from the government as the first generation does, these rights necessitate a proactive government acting on behalf
of its citizens. They establish an acceptable standard of living for all—that
is, a minimal level of equality.

The third generation of rights is referred to as *solidarity* rights since
their realization requires the cooperation of all countries. It claims rights
for those that were latecomers to the industrialization that swept the Western countries. They are the people of the South, or the third world, many
of whom until the past forty years lived under the burden of colonialism
and were not represented at the UN when the UDHR was approved. They
constitute 80 percent of the world's population but receive a very small portion of its benefits. The third generation of rights is a hope or even a demand for the global redistribution of opportunity and well-being. These
general rights are implied in Article 28: "Everyone is entitled to a social
and international order in which the rights and freedoms set forth in this
Declaration can be fully realized."

This generation does not have the status of other rights and is in the
process of being developed. Law professor Burns Weston says the following about them:

> [They] appear so far to embrace six claimed rights. . . . Three of these re-
> flect the emergence of Third World nationalism and its demand for a
> global redistribution of power, wealth, and other important values: the
> right to political, economic, social, and cultural self-determination; the
> right to economic and social development; and the right to participate in
> and benefit from "the common heritage of mankind" (shared earth-space
> resources; scientific, technical, and other information and progress; and
> cultural traditions, sites, and monuments). The other three third-generation
> rights—the right to peace, the right to a healthy and balanced environ-
> ment, and the right to humanitarian disaster relief—suggest the impotence
> or inefficiency of the nation-state in certain critical respects. (Weston
> 1992: 19–20)

As an example of the implementation of the third-generation solidarity
rights, the General Assembly adopted the Declaration on the Right to Development in December 1986. As Winston Langley notes,

> The Declaration confirms the view of the international community that the
> right to development is an inalienable human right "by virtue of which
> every human person and all peoples are entitled to participate in, con-
> tribute to and enjoy economic, social, cultural and political development,
> in which all human rights and fundamental freedoms can be fully real-
> ized." (Langley 1996: 361)

Some students of human rights merge the second and third generation of rights; the UN has done this in its *Human Development Report 2000: Human Rights and Human Development.* This volume, published annually by the United Nations Development Programme (UNDP), argues that human rights cannot be realized without human development, and human development cannot be realized without human rights.

The final two articles of the Universal Declaration of Human Rights affirm the universality of and responsibility for the rights described in Articles 1–28:

29. (1) Everyone has duties to the community in which alone the free and full development of his personality is possible.
    (2) In the exercise of his rights and freedoms, everyone shall be subject only to such limitations as are determined by law solely for the purpose of securing the recognition and respect for the rights and freedoms of others and of meeting the just requirements of morality, public order and the general welfare in a democratic society.
    (3) These rights and freedoms may in no case be exercised contrary to the purposes and principles of the United Nations.
30. Nothing in this Declaration may be interpreted as implying for any State, group or person any right to engage in any activity or to perform any act aimed at the destruction of any of the rights and freedoms set forth herein.

■ THE UN AND HUMAN RIGHTS IMPLEMENTATION

Tom Farer observes that the UN operates at four levels in supporting human rights (Farer 1992: 232–235). First, it formulates and defines international standards by approving conventions and making declarations. Second, the UN advances human rights by promoting knowledge and providing public support. At the third level, the UN supports human rights by protecting or implementing them. Although the task of directly enforcing human rights is primarily left to the states themselves, the UN does become involved in various means of implementation.

During the 1990s and early part of the 2000s, UN enforcement took on new meaning and controversy. The efforts by the UN, often under pressure from the United States and its allies, in the Persian Gulf, Somalia, Rwanda, the former Yugoslavia, and Afghanistan are examples of this. They include the boycotts of aggressor states, military action, military support for the delivery of humanitarian aid, and the protection of refugees. Some of these are controversial extensions of the UN mandate and will provide material for discussions about the role of the UN in the future. Finally, the UN has taken additional steps at enforcement that some consider to be structural and economic aspects of human rights issues. This refers to support for the third generation of rights, including economic development for poorer countries as described earlier. Development has taken a great deal of UN resources but does not receive the public attention that more dramatic actions do.

Beyond the positive actions that the UN takes in supporting human rights, there is a great deal of criticism of their inability to do more. Following are some of these criticisms: First, some countries have signed conventions but have made little progress toward instituting them. Why do countries sign these conventions with no apparent intention to enforce them? All countries of the world want to appear to other countries and persons as though they treat their citizens justly. Additionally, this interest in being perceived as just and humane to one's citizens can work to encourage the observation of human rights by such governments.

Second, some countries, regardless of their human rights record, are reluctant to subject their country to the jurisdiction of such world bodies. The United States, which generally has a good human rights record, often fits into this category. It seems that the United States refuses to ratify the Rights of the Child Convention (see Chapter 11), the founding of the International Criminal Court, and several other conventions because it wants to determine its own law. This is part of the ongoing discussion about state sovereignty and the UN. In short, a country says that what happens within its own borders is its business and not the concern of other political bodies.

A third reason for the failure of these approved human rights agreements is the structure of the UN itself. The UN agencies that are given the responsibilities for enforcement are not sufficiently independent and are not adequately funded.

Fourth, and related to some of the above criticisms, the UN is not an independent body or a world government; it is subject to the whims of its members and has no more power or resources than it is given by its members. For instance, many UN decisions on human rights violations are made in the Security Council with its five powerful permanent members—Britain, China, France, Russia, and the United States, any of which can veto an action. (The ten nonpermanent members do not have such power.) Thus, any action taken must be approved by all five members, a very difficult task. Another inhibiting factor is the availability of money for UN operations. Several countries are either unable or unwilling to pay their rightful share. Again, the United States comes to mind as it has been behind on its payments by some one billion dollars in recent years. In this case it is not a lack of money but rather political differences in the U.S. Congress over support for UN actions and policies that explains the unpaid bill.

■ HUMAN RIGHTS IMPLEMENTATION
OUTSIDE THE UNITED NATIONS

With all of its problems one cannot speak of human rights without invoking the name of the UN. It is here that human rights issues emerge and major discussions take place. With all of its weaknesses, UN efforts must be seen within the total context of human rights efforts. Because of the centrality of human rights issues in the world (greatly due to the work of the UN), governments and individuals have founded other institutions to supplement this very important work. These groups have taken different forms, received different degrees of approval, and seen varying degrees of success.

One non-UN force to enter the human rights arena is the regional human rights structure. The most advanced and effective of these is in Europe and operates under the Convention for the Protection of Human Rights and Fundamental Freedoms. It was established in 1950 and functions under the European Commission of Human Rights. The commission receives complaints from approximately 4,000 individuals per year. (Many human rights agencies will only receive complaints from governments.) After analysis, this large number is reduced to some forty cases per year. The commission "pursues [these] vigorously, and a majority end with a decision against the state" (Donnelly 1993: 82). Thus, "individual human rights" are realized, even though they may be opposed by the laws of sovereign states. According to human rights advocates, this sophisticated, well-funded, and very successful regional structure is a model for the whole world to emulate.

A similar, but much less successful structure exists in the Americas. It includes the seven-member Inter-American Commission of Human Rights and the Inter-American Court of Human Rights. Its decisions have often been resisted or ignored by countries of the Americas on the principle that state sovereignty predominates. The commission and court are left with the power of publicity and moral influence, which have been quite limited.

Finally, in the 1980s, African states approved the African Charter on Human and Peoples' Rights. It is an interesting document in that, unlike other regional documents, it includes the rights of "peoples," that is, the third-generation, or solidarity, rights. Article 19 states, "All peoples shall be equal; they shall enjoy the same respect and shall have the same rights. Nothing shall justify the domination of a people by another." The document does not provide for a human rights court but emphasizes mediation, conciliation, and consensus.

A second and very promising human rights development outside of the UN has been the activities of nongovernmental organizations (NGOs). Since many countries seem to have limited commitment to human rights, NGOs have come to serve the purpose of monitoring human rights violations throughout the world. Hundreds of these organizations exist, but again they are more active and successful in areas where general support for human rights already exists. In several countries where human rights have made little headway, the very presence of NGOs is often unsupported, if not prohibited, and their advocates and employees prosecuted. In many parts of the world it is dangerous to work for human rights groups.

NGOs take many forms and operate in many different ways. By operating outside of government they are able to monitor the actions of governments and bring pressure on governmental policies. Laurie Wiseberg lists nine areas in which human rights NGOs provide services: (1) information gathering, evaluation, and dissemination; (2) advocacy to stop abuses and secure redress; (3) legal aid, scientific expertise, and humanitarian assistance; (4) national and international lobbying; (5) legislation to incorporate or develop human rights standards; (6) education, conscientization (raising the consciousness of citizens), or empowerment; (7) solidarity building; (8) delivery of services; and, (9) access to the political system (Wiseberg 1992: 73–77).

Amnesty International, Human Rights Watch, the International League for Human Rights, Cultural Survival, the International Commission of Jurists, the International Committee of the Red Cross, Physicians for Human Rights, and many other NGOs provide a variety of human rights services. Although they are somewhat limited in what they can do, they represent one of the most promising avenues for human rights support. The UN has made provisions for these organizations to have an official representation at the UN and at UN-sponsored conferences. Geoffrey Robinson, a scholar who has been quite critical of the UN, argues that experts from human rights NGOs should be eligible for appointment to UN committees and commissions:

> The best way forward is to bring non-government organizations (which do most of the real human rights fact-finding) into the appointments process, thereby providing some guarantee that members are true experts in human rights, rather than experts in defending governments accused of violating them. (Robinson 2000: 47)

■ CURRENT ISSUES RELATED TO HUMAN RIGHTS

The emergence of human rights on the world's political agenda in recent decades is clearly not without controversy. At every step since the signing of the Universal Declaration of Human Rights, there have been delays and denouncements. There is no reason to believe that this will stop. In the sections that follow, I address a few of these controversies and the arguments that surround them.

The UN Charter guarantees state sovereignty or self-determination and nonintervention; it also states that all individuals, regardless of their citizenship and status, have human rights. These principles are often found to be contradictory. The idea that everyone possesses these rights as found in the UDHR is referred to as universalism. On the other hand, some countries and cultures follow traditions that are considered inconsistent with the UDHR, and they claim exception for their traditions. These governments say they are the final authority in determining what is right for their citizens: they plead state sovereignty, which has been a long-standing global principle for some three centuries. According to these countries, appropriate expectations for human rights are judged against, or relative to, local culture. In other words, certain acts thought by some to violate human rights might be upheld as legitimate cultural or religious practices by others. This view is referred to as relativism or cultural relativism. Two such acts or customs are child brides and female circumcision (see following section). In South Asia young girls are often promised by their families to marry at an early age and without considering the desires of the child. This is generally considered a violation of the child's rights, but it is often defended as a cultural tradition, or relativism. For those who participate in this practice the determining factor would be tradition, not an abstract rule that is considered to apply universally.

As the debate proceeds over what rights are universal, most decisions will fall somewhere between these two extremes of universalism, which states that everyone has certain rights regardless of the circumstances, and relativism, which holds to the idea that local customs and culture determine what rights people have. At the same time, however, there is general agreement that such acts as genocide (the killing of a people such as Jews, Gypsies, Hutus, or Tutsis), torture, and summary executions are violations of human rights.

■ Female Circumcision

Female circumcision (sometimes called female genital mutilation, or FGM) is a cultural practice that has become a high-profile issue and is defended as a tradition by some and condemned by others as a violation of human rights. This procedure may involve the complete removal of the clitoris and occasionally the removal of some of the inner and outer labia. In its most extreme form—infibulation—almost all of the external genitalia are cut away, the remaining flesh from the outer labia is sewn together, or infibulated, and the girl's legs are bound from ankle to waist for several weeks while scar tissue closes up the vagina almost completely. A small hole, typically about the diameter of a pencil, is left for urination and menstruation (McCarthy 1996: 32).

Source: Jim Borgman, *Cincinnati Enquirer*, 1996. Reprinted with special permission of King Features Syndicate. © Copyright King Features Syndicate.

It is estimated that this procedure affects some 137 million women, mostly in Africa. The cultural reasons for supporting it are that it makes girls "marriageable" (because it ensures their virginity) and also diminishes their sex drives. It is viewed by its proponents as a cultural practice that is a personal and local matter that should not be considered a human rights violation. Yet a majority of the UN members saw things differently and in 1993, in the Declaration on the Elimination of Violence Against Women, Article 2a, condemned

> physical, sexual, and psychological violence occurring in the family, including battering, sexual abuse of female children in the household, dowry-related violence, marital rape, female genital mutilation, and other traditional practices harmful to women. (Langley 1996: 1606)

If it were not for the international discussion of FGM, it is doubtful that Fauziya Kasinga, a native of Togo, would have received asylum in the United States after a year in a U.S. prison. After coming to the United States and admitting to immigration officials that she did not have a valid passport, it took a year of intense legal effort before the U.S. Board of Immigration Appeals granted political asylum to Kasinga, recognizing female circumcision as a form of persecution against women. The ruling sets a binding precedent for all U.S. immigration judges. It also leaves open the possibility that women threatened with this practice may seek asylum in the future (Burstyn 1995: 16).

■ State Sovereignty and the Question of Impunity

State sovereignty is a cornerstone of the present international system. But like so many things in our globalizing world it is being challenged and slowly altered. For instance, at the close of the Gulf War in April 1991, the Security Council passed Resolution 688, which permitted the establishment of temporary havens for refugees inside Iraq and without the permission of Iraq. The rationale was that the violent treatment of Kurds (a large ethnic group living in Iraq) by the government of Iraq threatened international peace and security. This clearly contradicts the traditional understanding of state sovereignty. A more recent example was the establishment of tribunals (or courts) to try persons responsible for crimes against humanity in the former Yugoslavia and Rwanda.

State sovereignty seems to have its corollary in the impunity extended to the heads of state, even those who perpetuate massive human rights abuses. Those that have killed tens of thousands and even millions of their own citizens—such as Idi Amin in Uganda, Augusto Pinochet in Chile, Pol Pot in Cambodia, and Slobodan Milosevic in the former Yugoslavia—have rarely been held responsible for their crimes. They have not gone unnoticed, but because of the Cold War and the fact that indicting them would confront state sovereignty, they have received limited attention. Yet these kinds of violations have not always been overlooked.

Crimes that were labeled crimes against humanity were dealt with after World War II in both Germany and Japan. Those war criminals were tried before tribunals and found guilty. Since then, however, there has been a long lull in much human rights work that lasted from 1946 to 1976, a period that Robinson refers to as "thirty inglorious years" (Robinson 2000: 37). During that time, the United States supported many governments that were responsible for massive human rights violations (crimes against humanity) but justified them on the basis of fighting communism. The USSR did the same thing with different justification—fighting Western imperialism and capitalism.

Thus, these men and those who supported them and carried out their grisly commands have generally been able to walk away from their crimes with little fear of being held responsible. Often, before leaving their posts as heads of state they would see that legislation was passed that would absolve them and their cohorts of any guilt—they could leave with impunity.

A crack in the armor of impunity (exemptions from responsibility for these acts) began to occur with the establishment of "truth commissions" in the mid-1970s to investigate some of these massive crimes against humanity. The first one was commissioned in Uganda in 1974 but was disbanded before it was completed. This was followed by twenty additional commissions that had varying degrees of success in exposing crimes. The aims of these truth commissions were fivefold: to discover, clarify, and formally acknowledge past abuses; to respond to specific needs of victims; to contribute to justice and accountability; to outline institutional responsibility and recommend reforms; and to promote reconciliation and reduce conflict over the past (Hayner 2001: 24–31). These reports helped to rekindle the interest in dealing with crimes against humanity as was done in the Nuremberg Trials after World War II.

Finally, as mentioned previously, in early 1993 and again in late 1994, two international tribunals were established by the UN Security Council to deal with crimes against humanity in the former Yugoslavia and in Rwanda in Central Africa. In the former Yugoslavia, some 200,000 people were killed in what was called ethnic cleansing (separating ethnic groups by killing or forced migration), and in Rwanda, approximately 800,000 were massacred in what appeared to be tribal violence.

The timing of these tribunals is especially important since they were established before the end of the conflicts and thus constitute a form of early intervention. Also, these tribunals were established based on international law, which supersedes state sovereignty, at least in principle. Observer David Scheffer argues that the UN Security Council establishment of these two tribunals under the authority of Chapter VII of the UN Charter set an important precedent. International humanitarian law can subsequently override domestic jurisdiction (that is, state sovereignty in such matters is challenged).

Challenges to state sovereignty have continued. On May 11, 1997, Dusto Tadic, a Bosnian Serb, was found guilty by the International Criminal Tribunal for the Former Yugoslavia (ICTY) of eleven counts of crimes against humanity. As Robinson notes, "This man will go down in history as the person whose case settled the principles and scope of international criminal law at the end of the twentieth century" (Robinson 2000: 310). But this is not the end; Tadic was an underling, one of thousands who willingly and joyfully tortured, raped, starved, and killed whomever they chose. What will come of those who mandated and oversaw these crimes?

The case of one leader began to unfold in the late 1990s, that of Augusto Pinochet, the military dictator who had overthrown the democratically elected government of Chile in 1973 and is presumed responsible for the disappearance and death of thousands of Chileans. Pinochet granted himself self-amnesty and was made senator for life; both circumstances gave him immunity from prosecution in Chile and presumably elsewhere in the world.

When Pinochet traveled to London for medical treatment in 1998, however, the Spanish government asked that he be extradited to Spain where he would be tried. After Pinochet made several court appearances in London, it was decided that he could be extradited to Spain to face charges. According to Robinson, "The Pinochet Case was momentous because—for the first time—sovereign immunity was not allowed to become sovereign impunity. The great play of sovereignty, with all its pomp and panoply, can now be seen for what it hides: a posturing troupe of human actors, who when off-stage are sometimes prone to rape the chorus" (2000: 374). Ultimately, because of Pinochet's age and health (he is said to have suffered a heart attack), he was permitted to return to Chile, but another door had opened to world justice.

The dramatic developments in the Tadic and Pinochet cases are outdone by the more recent case of Slobodan Milosevic, who was defeated as president of Yugoslavia in a national election there in 2000. To the surprise of many, the new Yugoslav government extradited him to stand trial at the ICTY, where he was indicted in 1999 but not arrested. Then in November 2001 he was indicted on charges of crimes against humanity in Croatia in 1991–1992 and Kosovo in 1999 as well as for genocide, crimes against humanity, and grave breaches of the Geneva Convention (*New York Times* 2001). This is a dramatic change on the human rights scene. It is a clear warning to heads of state that just because they occupy a premier position they are not immune from international prosecution if they commit crimes against their own people.

■ Human Rights: Western Imperialism?

Does the growing influence of human rights on the world scene constitute a form of cultural imperialism? This accusation is heard from several East Asian countries—China, Indonesia, and Singapore among them—who view human rights as a product of Western civilization and not fully applicable to their societies. Although they do not refer to what are considered the more serious violations of human rights—such as murder, slavery, torture, and genocide—as controversial, they do take issue with first-generation rights such as freedom of the press, speech, association, and expression. They argue that they had little input into the Universal Declaration of Human Rights and that it expresses values they do not necessarily support (Bell 1996).

One argument these poorer Asian countries give to support their position is that economic development necessitates at least the temporary suspension of some rights. They say that suspending certain rights will result in a greater good for more persons in the long run, when economic growth is realized. Western supporters of human rights respond that there is little evidence that human rights inhibit economic growth. Some argue that quite the opposite is true—that social and political rights may help ensure such economic growth.

East Asian countries also argue that Western human rights advocates overlook the negative consequences that the Western emphasis on individual freedom brings. They point to the many social problems that the West, especially the United States, is experiencing. Such problems as drug abuse, crime, declining commitment to the family, homelessness, racism, and general alienation feed this skepticism. These countries feel that the focus on first-generation rights as opposed to second-generation rights is at least partially responsible for Western moral decline. Further, they argue that the unjustifiable U.S. involvement in the Vietnam War and close U.S. relationships with and support for many Asian governments that have massively violated human rights are more reasons to be skeptical of Western promotion of human rights (Faison 1997).

The recent controversy over human rights in China is directly related to the issue of trade. China has long sought greater access to global markets. For China this opening could come through membership in the World Trade Organization (WTO), yet this had been held up for some fifteen years because of China's human rights record. In 1989, the Chinese government broke up a prodemocracy student movement in Tiananmen Square with ruthless disregard for life and was broadly condemned by the West. This was added to the charges that in China arbitrary arrests are common and that the Chinese government disregards several other first-generation rights. Add to these the fact that China is not a democracy and reportedly sells body parts of recently executed prisoners. Should a country with such a human rights record be permitted to benefit from membership in the WTO?

For several years, the United States has cosponsored a resolution at the UN Human Rights Commission in Geneva to call for an investigation of China's human rights record. On all occasions China vetoed the resolution. In the late 1990s things began to change. China made some human rights concessions and in 1996 the United States established some conditions under which the resolution would not be reintroduced. China has released many political prisoners and incorporated additional civil and political human rights concepts into its laws.

In November 2000 China and the UN did sign a human rights pact. Mary Robinson, the UN high commissioner for human rights, and Chinese vice foreign minister Wang Guangya "signed a memorandum committing China to comply with rights treaties it has already signed and to review some current rights abuses, including its use of labor camps" (*New York Times* 2000).

Whether or not significant human rights changes are taking place in China is still debatable. The question seems to focus on whether human rights changes in China would come about if China were permitted to become a full partner in the global economy or whether the chance of changes would be greater if they were kept out. After years of debate the United States and the European Union agreed that permitting China to join the WTO was the appropriate thing to do, and in late 2001 that happened. Only time will tell how this decision will play out.

■ CONCLUSION

Human rights have become an integral part of the international political landscape and, if present trends continue, will become even more important in the future. This trend, however, will not necessarily continue. The terrorism that the United States experienced in September 2001 could alter this trajectory as could the response to that terrorism. If the terrorism continues, it could disrupt the international cooperation essential to increased human rights. Similarly, if in its response the United States disregards the legal processes essential to the rule of international law, this would also impede human rights. On the other hand the post–September 11 period presents the possibility that greater cooperation will take place on global issues, such as human rights.

Maintaining world order is a delicate balancing act with uncertain outcomes. Instituting and maintaining human rights in the international community has come to be seen as one way of incorporating regularity and predictability in our daily lives. The next few years will be crucial in determining whether or not the human rights cause can continue to play that role.

■ DISCUSSION QUESTIONS

1. Which generation of human rights do you think is most important?
2. Why does the UN not enforce the human rights that the General Assembly has already approved?
3. Should such practices as female circumcision be seen in universal or relative terms?
4. Which is most important: state sovereignty or universal human rights?
5. Should China have been permitted to benefit from joining the World Trade Organization in light of their human rights record?

■ SUGGESTED READING

Amnesty International Report (annual) London: Amnesty International Publications.
Claude, Richard Pierre, and Burns H. Weston, eds. (1992) *Human Rights in the World Community.* Philadelphia: University of Pennsylvania Press.

Donnelly, Jack (1993) *International Human Rights*. Boulder, CO: Westview Press.

Felice, William F. (1996) *Taking Suffering Seriously*. Albany: State University of New York Press.

Forsythe, David P. (1991) *The Internationalization of Human Rights*. Lexington, MA: Lexington Books.

Gutman, Roy (1993) *A Witness to Genocide*. New York: Macmillan.

Hayner, Priscilla B. (2001) *Unspeakable Truths: Confronting State Terror and Atrocity*. New York and London: Routledge.

Human Rights Watch World Report (annual) New York: Human Rights Watch.

Langley, Winston E. (1996) *Encyclopedia of Human Rights Issues Since 1945*. Westport, CO: Greenwood Press.

New York Times (2000) "United Nations, China Sign Human-Rights Pact." November 20.

Robinson, Geoffrey (2000) *Crimes Against Humanity: The Struggle for Global Justice*. New York: W. W. Norton.

Staub, Ervin (1989) *The Roots of Evil: The Origins of Genocide and Other Group Violence*. Cambridge: Cambridge University Press.

5

PEACEKEEPING
AND PEACEMAKING

Carolyn M. Stephenson

Conflict now routinely crosses international boundaries, blurring the distinction between domestic and international conflict. In response, new methods of providing for international peace and security have evolved, especially in the post–Cold War world since the early 1990s. Peacekeeping and peacemaking have taken new forms, and the role of multilateral institutions, especially the United Nations (UN), has become much more significant. What was initially optimism for a more peaceful world in the wake of the Cold War, however, has hardened into a recognition that we still face conflicts as intractable as the Cold War between the Soviet and Western blocs appeared to be from the late 1940s to the late 1980s.

In the late 1980s, the world was full of hope for a new, more gentle order that would provide for more peace and security. There were hopes for a renewed UN, for new forms of mediation and other third-party conflict resolution, for a new relationship between the superpowers, for the signing of arms control and disarmament treaties, and for the increasing recognition of individual human rights and needs. Nonviolent revolutions had overturned authoritarian regimes in the Philippines in 1986 and in Eastern Europe in 1989. The UN had received the Nobel Peace Prize for its peacekeeping missions in 1988. We were beginning to develop solid international agreements to protect and restore our environment.

We were less hopeful and more sober by the early 1990s. We had begun to take seriously both new dimensions of conflict and new approaches to peacemaking, yet disagreed about what constituted the grounds for successful peacemaking. For some, success consisted of the breakdown of the Soviet Union and the renewed ability of the United Nations to function as originally intended as maker and keeper of the peace, under the leadership

71

of the United States. Many saw U.S. military, economic, and political power as having been responsible for the end of the Cold War. For them, the restoration of the UN capability for enforcement, under U.S. leadership, was central.

For others, what constituted success was the restoration of a different United Nations, a United Nations that would run by one-nation, one-vote, countering the dominant influence of both superpowers. This UN would function cooperatively to further individual and group rights, security, development, and the state of the global environment. For these people, U.S. leadership was not so important as that of decentralized political movements of individual human beings. For them, the changes of the 1980s had come about not so much because of U.S. military, economic, and political power but because of the committed organizing power of social movements all over the world.

The two approaches, and other variants of them, rely on different conceptions of security and peace and on different conceptions of power. Because they emphasize different methods of and approaches to peacemaking, it is important that we examine the conceptions of security, peace, and power that underlie these approaches. Otherwise, we risk shifting from notions of world government to arbitration, to nuclear deterrence, to rapid deployment forces, to UN peacekeeping, to nonviolent revolution, to arms control, to mediation, or to humanitarian intervention, with no sense of why we have shifted from one approach to another, let alone what the strengths and weaknesses of each approach are in particular situations.

■ CHANGES IN THE CONCEPT OF SECURITY

Where national security was once virtually the only way to talk about security, the world has come to acknowledge the relationship between national security and both international and individual security. We have moved from reliance on a balance-of-power system, to collective security, collective defense, and then common security, with the present international security system representing some mixture of all of these.

The classical *balance-of-power system*, the primary system for maintaining security in nineteenth-century Europe, was retained well into the twentieth century. With a goal of ensuring that no nation-state became so strong as to be able to overpower others, rough equality was maintained by two camps of states in the system, with one or several states (usually Britain) changing alliances in order to maintain the balance. This system began to break down in the twentieth century, when it failed to avert war and maintain stability.

Under *collective security*, which began with the League of Nations in 1919 and was strengthened with the establishment of the United Nations in 1945, states agreed on certain rough rules of international law, including national sovereignty and freedom from outside aggression; they also agreed that if any state violated these rules, all of the others would band together against that state. Sanctions for violating the prohibition against international aggression could be either military or nonmilitary. The UN Security Council is the primary site of collective security, with Chapter VII of the UN Charter governing this.

Collective defense, which was a step back in the direction of the balance-of-power system, became the dominant security system by the late 1940s and throughout the Cold War. Under this system, each set of nation-states, West and East, gathered together in military alliances to defend against the other set. The West embraced capitalism and democracy, while the East promoted communism, each hoping to establish its systems throughout the rest of the world. The formation of the North Atlantic Treaty Organization (NATO) by Western states in 1949 was followed by the formation of the Warsaw Pact by Soviet-bloc states in 1955. Each side bolstered its conventional military defenses with the nuclear umbrella of its respective superpower. Although collective defense has not entirely ended, the withdrawal of Soviet forces from Eastern Europe in the late 1980s, together with the end of the Warsaw Pact in 1990, left NATO with questions about its remaining purpose. Deterrence, including but not limited to nuclear deterrence, was the primary underlying power dynamic of collective defense.

In contrast, the concept of *common security* arose, primarily within the UN framework. There are two distinct aspects of common security, one of which arose in the context of North-South conflict, one in the East-West context. With respect to the first, the Report of the Independent Commission on International Development Issues in 1980, better known as the Brandt Report, raised notions of economic security (ICIDI 1980). For the South, the failure of economic development was perceived as a much greater threat to security than nuclear war. In 1982 the Independent Commission on Disarmament and Security Issues, or the Palme Commission, made two more explicit linkages: first, there could be no victory in nuclear war—therefore we could only survive together; second, the costs of militaries everywhere were contributing to economic insecurity—therefore the reduction of military costs could contribute to development (ICDSI 1982).

The East-West aspect of common security is embodied in the Helsinki Agreement and is the best known example of common security. This agreement (today called the Organization for Security and Cooperation in Europe) was formed in 1975 by the states of both NATO and the Warsaw Pact and contains three "baskets": a security basket, which includes agreements on post–World War II borders in Europe; an economic basket, which opens up trade between East and West; and a human rights basket, which provides for certain human rights guarantees and procedures in both East and West.

The Brundtland Commission report, *Our Common Future*, in 1987 added the concept of *environmental security* to that of common security, strengthening the idea that sustainable development requires sustaining the environment that supports development and promoting the notion that military expenditures and war are harmful to the environment. Environmental security encompasses both the protection of the environment for its own sake and the protection of the environment for the sake of humankind. Thus, common security now includes political-military aspects, economic aspects, and environmental aspects and seems to be recognizing their interdependence; however, it is even more explicit in acknowledging the interdependence of states with regard to security.

Security today is thus conceived of in a much more comprehensive way than ever before, even when that security applies still to the nation-state rather than global society. The concept comprises not only negative security (the ability to defend against or shut off relationships one views as harmful) but also positive security (the ability to maintain relationships one views as essential to one's survival, such as access to food, oil, and credit). Such a reconceptualization of security means that reliance on traditional approaches to security are less likely to be adequate. This is one of the reasons new approaches to peacemaking are being taken ever more seriously.

■ CHANGES IN THE CONCEPT OF PEACE

The concept of peace has also broadened in much the same way as security, expanding from the concept of negative peace alone—or peace as the absence of war or direct violence—to include positive peace—or peace as the absence of exploitation and the presence of social justice. While the earliest mention of positive and negative peace appears to be in the writings of Martin Luther King, Jr., the terms were expanded upon and more fully analyzed and operationalized by Johan Galtung (1969). The debate that ensued over which concept of peace was to be accepted has yet to be resolved. For some, the absence of direct violence seems more important; for others, the absence of exploitation is key. But until we are more in agreement about the kind of peace we are interested in making, there will continue to be major differences among approaches to peacemaking.

At one end of the spectrum, those who believe that peace is simply order or the absence of violence may argue that peace can or should be enforced with military action. At the other end of the spectrum, those who believe that peace must include justice may argue that peace cannot be enforced and can only be brought about by negotiations that take into account justice and the underlying needs of the parties. The international system has mechanisms that span the full range of these points of view.

For a long time, distinctions have been made in the UN between peace-making, peacekeeping, and peace-building (Boutros-Ghali 1992). Without getting into an argument over technical definitions, let it suffice to say that peace-building generally includes building the conditions of society so that there will be peace. In this area we might include such methods as human rights education, development and development aid, and reconciliation and the restoration of community following a violent conflict. Peacekeeping, in the broader sense, involves keeping parties from fighting or otherwise doing harm to each other. In the narrower sense, it has been used to describe the particular multinational operations employed to restore and maintain peace between hostile parties. Peacemaking is usually taken to mean helping to bring hostile parties to agreement. I will explore all three of these, with an emphasis on peacekeeping and peacemaking and with the recognition that they overlap somewhat.

■ PEACEMAKING AND THE UNITED NATIONS CHARTER

The UN Charter includes two primary ways of providing for peacemaking and for international peace and security. Chapter VI of the Charter focuses on Peaceful Settlement of Disputes, while Chapter VII relates to Action with Respect to Threats to the Peace, Breaches of the Peace, and Acts of Aggression. Article 33, the first article of Chapter VI, provides that

> the parties to any dispute, the continuance of which is likely to endanger the maintenance of international peace and security, shall, first of all, seek a solution by negotiation, enquiry, mediation, conciliation, arbitration, judicial settlement, resort to regional agencies or arrangements, or other peaceful means of their own choice.

When parties are unable to negotiate their way through a dispute on their own, the Security Council may call upon them to settle their dispute by any of these means, and it, or other parts of the United Nations, may assist them by providing a third party to help them do so. *Enquiry and fact-finding* are methods by which a third party attempts simply to find out the facts of the situation. In *mediation and conciliation,* a third party, sometimes in the form of a special representative of the Secretary-General, sometimes in the form of a commission, helps the conflicting parties communicate and come to agreement when they are unable to do so. Under *arbitration*, the third party makes a decision about the conflicting claims of the parties; sometimes the parties can choose whether to accept this decision, but under binding arbitration, the parties are bound to accept the decision of the arbitrator. Under *judicial settlement*, the Charter provides for submitting certain types of legal disputes between states to the International

Court of Justice in The Hague, Netherlands, where the court will rule on the legitimacy of the case under international law.

Mediation has become increasingly important in resolving international disputes. The "quiet diplomacy" of the UN Secretary-General or his representatives has been used in a long series of crises in the Middle East, beginning in the late 1940s and continuing through to the Iran-Iraq War, as well as in other regional areas. Mediation has also been used by international regional organizations such as the Organization for African Unity, as well as by powerful states such as the United States—for example, when it mediated the peace between Israel and Egypt during the Camp David negotiations—and by less powerful states such as Algeria, which mediated the release of the U.S. hostages held by Iran. Mediation is also practiced in the form of what is called second-track diplomacy, where individuals such as academic specialists in conflict resolution or representatives of the International Committee of the Red Cross or of religious organizations, such as the Mennonite Conciliation Service or the Friends World Committee for Consultation (Quakers), help to facilitate communication or to run workshops aimed at solving the problems underlying the conflict.

UN enforcement action constitutes another approach to peacemaking that has become more available to the international system since the end of the Cold War. Enforcement action is covered under the collective security provisions of Chapter VII of the UN Charter, especially the *nonmilitary sanctions* provided in Article 41 and the *military sanctions* provided in Article 42. While some point out that the Charter conditions for military enforcement have never been met, due to the failure to set up UN forces under a Military Staff Committee (Articles 43–47), most would agree that UN-authorized operations in Korea in 1950 and in Iraq and Kuwait in 1991 constitute the primary examples of UN military enforcement. Enforcement action is generally considered when a state has clearly violated the terms of the Charter and carried out aggressive international action. Military enforcement actions have tended to be led and staffed by one or several of the great powers. There are clear differences of opinion as to whether military enforcement action constitutes an approach to peacemaking or is better considered simply as war.

Nonmilitary sanctions are another approach to peacemaking. Article 41 of the UN Charter says that "the Security Council may decide what measures not involving the use of armed force are to be employed to give effect to its decisions," including "complete or partial interruption of economic relations and of rail, sea, air, postal, telegraphic, radio, and other means of communications, and the severance of diplomatic relations."

The old debate over whether sanctions are appropriate and effective was renewed after Iraq's invasion of Kuwait. Peace organizations generally supported sanctions as an alternative to war before the war, but opposed sanctions as harmful to the Iraqi people after the war. If the purpose was to get Saddam Hussein out of Kuwait, there were early indications that sanctions might have worked. If the purpose, on the other hand, was to get Saddam Hussein out of Iraq, sanctions were not likely to be effective. The ambiguity between those two goals in messages from the United States may be one of the reasons sanctions did not succeed within the time period they were used.

In the 1930s, sanctions were seen as a primary guarantor, within the system of collective security, for preventing wars. However, after the failure of the League of Nations sanctions against Italy in 1936 with respect to Ethiopia, sentiment turned against sanctions. Most observers today have concluded that sanctions are not especially useful on major foreign policy goals but carefully focused sanctions may be useful in limited goals.

The definition of *sanctions* has changed over time. By the time of the League of Nations, sanctions meant actions taken by international bodies to enforce international law. Since then, the term has come to include unilateral acts and even the use of economic policies for ordinary diplomatic influence. Evaluation of the success of sanctions may be very different if one separates out the unilateral from the more consensually based actions of international organizations.

■ PEACEKEEPING

On October 10, 1988, UN Secretary-General Javier Pérez de Cuéllar accepted the Nobel Peace Prize on behalf of the 10,500 members of UN peacekeeping forces. He paid tribute to the half-million young men and women from fifty-eight countries who served in UN peacekeeping operations, especially to the 733 who had lost their lives. Both he and Norway's prime minister, Gro Harlem Brundtland, expressed their concern for the financial status of peacekeeping, particularly the fact that major powers are in considerable arrears. At that time, the annual cost of peacekeeping activities was about $300 million. Annual UN peacekeeping expenditures peaked at $3.6 billion in 1993, while the number of missions peaked at twenty-one in 1999. Peacekeeping had come of age but was endangered by the massive costs that accompany certain of its activities.

There is no official UN definition of *peacekeeping*. However, the definition adopted by the International Peace Academy (IPA), a nongovernmental organization (NGO) closely related to the UN that has undertaken much of the training for UN peacekeeping forces, has been seen as close to official. Under that definition, peacekeeping is

> the prevention, containment, moderation and termination of hostilities
> between or within states, through the medium of a peaceful third party

intervention organized and directed internationally, using multinational forces of soldiers, police, and civilians to restore and maintain peace. (IPA 1984: 22)

A wide range of interpretations are still possible under that definition. UN publications distinguish between two kinds of peacekeeping operations: observer missions and peacekeeping forces. Observers are not armed; soldiers in peacekeeping forces have weapons but generally are authorized to use them only in self-defense. By 1990 there had been ten observer missions and eight peacekeeping forces. By June 1996 there had been forty-one peacekeeping operations (UNDPI 1990: 3; UNDPI 1996), by 2001 fifty-four. Peacekeeping is not based on sending a fighting force to stop a violent conflict. Rather, the premise of peacekeeping is that inserting an impartial presence in the region will allow the parties to try to negotiate a peaceful settlement to the conflict. There are differences in emphasis between the military and civilian role in peacekeeping operations—and different proportions of military and civilians in each operation—to achieve the double objective of reducing the violence and helping to move toward peaceful settlement.

Peacekeeping is clearly different from enforcement action or action based on collective security. The separation of peacekeeping from enforcement is critical as both peacekeeping and enforcement are made increasingly possible by the condominium of the great powers in the aftermath of the Cold War. The fact that peacekeeping missions are deployed in countries with their consent, that they are unarmed or lightly armed and use force only in self-defense, and that they are composed largely of middle-level powers whose degree of neutrality in the conflict is likely to be perceived as higher than that of the great powers, are significant factors that may well be important to their success.

The changing world situation in the late 1980s led not only to an increase in peacekeeping but also to the first UN Security Council summit meeting. In the concluding statement of the summit on January 31, 1992, the heads of state invited the Secretary-General to prepare "an analysis and recommendations on ways of strengthening and making more efficient within the framework and provisions of the Charter the capacity of the United Nations for preventive diplomacy, for peacemaking and for peace-keeping" (Boutros-Ghali 1995: 117–118). Secretary-General Boutros Boutros-Ghali, in his resulting report, spoke of the increasing demands for peacekeeping:

Thirteen peace-keeping operations were established between the years 1945 and 1987; 13 others since then. An estimated 528,000 military, police and civilian personnel had served under the flag of the United Nations until January 1992. . . . The costs of these operations have aggregated some $8.3 billion till 1992. . . . Peace-keeping operations approved at present are estimated to cost close to $3 billion in the current 12-month period. (Boutros-Ghali 1995: 57–58)

A mechanism not originally provided in the UN Charter had evolved over time to become the major item in the UN budget, with the exception of the specialized agencies.

■ PEACEMAKING AND PEACEKEEPING AFTER THE COLD WAR

At the end of the Cold War, when the focus of conflict turned from the threat of nuclear war to ethnic conflict in the developing world and elsewhere, there was an enormous rise in peacekeeping operations. Figure 5.1 indicates the geographic spread of both completed and ongoing UN peacekeeping operations from 1948 to 2001.

Table 5.1 shows that the fifteen current operations vary greatly in start date, size, budget, and purpose, as is revealed to some degree in the title of each mission. While a few of the current operations are almost as old as the UN itself, most began in the 1990s.

Many described these conflicts as a new type, although in reality they were not significantly different from the hundreds of intrastate ethnic, religious, racial, tribal, and national conflicts that had been occurring with regularity throughout the Cold War. The difference was that throughout the Cold War period the two sides had interpreted these conflicts as involving competition between communism and capitalism, while now they were seen simply as intrastate conflicts.

The United Nations Security Council, consisting of fifteen of the 189 members of the UN, votes by a 9-of-15 majority vote. The vote must not include a veto of any of the permanent members (China, France, Russia, the United Kingdom, and the United States). At the end of the Cold War and its pattern of Cold War vetoes, the Security Council found that it could make decisions and take action it could not have taken before. This led not only to a tremendous rise in the number of peacekeeping operations but also to a blending of traditional peacekeeping with enforcement in some cases and with peace-building measures such as election monitoring, human rights monitoring, and education, among other techniques, in others. These new types of operations sometimes came to be called "second-generation" peacekeeping. Missions such as those in Namibia, Cambodia, El Salvador, Haiti, and Mozambique are widely regarded as successes; they helped oversee elections and rebuild societies where there was agreement to do so. On the other hand, in places suffering a breakdown of society, including rioting

Figure 5.1 UN Peacekeeping Missions, 1948–2001

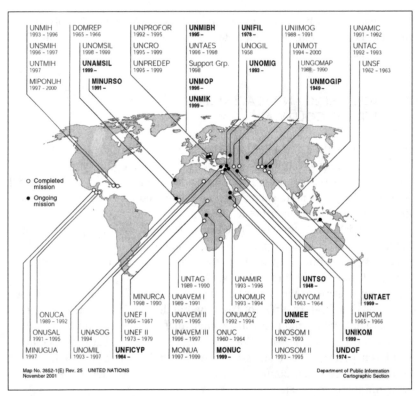

Map No. 3852-1(E) Rev. 25   UNITED NATIONS
November 2001

Department of Public Information
Cartographic Section

Source: United Nations Department of Public Information, Peace and Security Section, in consultation with the Department of Peacekeeping Operations and the Peacekeeping Financing Division, Office of Programming Planning, Budget and Accounts. DPI/1634/Rev.21a, September 15, 2001. Used by permission of the United Nations.

Notes: Peacekeeping operations 1948–September 15, 2001 total 54.

Operations under way on September 15, 2001 total 15.

Personnel as of August 31, 2001: Military personnel and civilian police serving total 47,151; countries contributing military personnel and civilian police total 88; international civilian personnel total 4,253; local civilian personnel total 8,489; total number of fatalities in peacekeeping operations 1948–September 15, 2001 total 1,680. The term *military personnel* refers to military observers and/or troops, as applicable. Fatality figures include military, civilian police, and civilian international and local personnel.

Financial aspects: Projection of costs from July 1, 2001–June 30, 2002 total about $3–3.5 billion; approved budget for period from July 1, 2000–June 30, 2001 total about $2.63 billion; estimated total cost of operations from 1948–June 30, 2001 total about $23.3 billion; outstanding contributions to peacekeeping on August 31, 2001 total about $3.4 billion.

Table 5.1 Current UN Peacekeeping Operations, 2002

UNTSO
UN Truce Supervision Organization
Since June 1948
Strength: military 153; international civilian
102; local civilian 109
Fatalities: 38
Appropriation for 2001: $22.8 million

UNMOGIP
UN Military Observer Group in India and
Pakistan
Since January 1949
Strength: military 45; international civilian
24; local civilian 42
Fatalities: 9
Appropriation for 2001: $7.3 million

UNFICYP
UN Peacekeeping Force in Cyprus
Since March 1964
Strength: military 1,251; international
civilian 41; civilian police 35; local
civilian 104
Fatalities: 170
Appropriation for July 2001–June 2002:
$42.4 million (gross), including
voluntary contributions of $13.6 million
from Cyprus and $6.5 million from
Greece

UNDOF
UN Disengagement Observer Force
Since June 1974
Strength: military 1,039; international
civilian 33; local civilian 88
Fatalities: 40
Appropriation for July 2001–June 2002:
$35.7 million (gross)

UNIFIL
UN Interim Force in Lebanon
Since March 1978
Strength: military 4,486; international
civilian 131; local civilian 340
Fatalities: 244
Commitment authority for July
2001–December 2001 (six months):
$106.2 million (gross)[a]

UNIKOM
UN Iraq-Kuwait Observation Mission
Since April 1991
Strength: military 1,099; international
civilian 57; local civilian 165
Fatalities: 14
Appropriation for July 2001–June 2002:
$52.8 million (gross), including voluntary
contributions of $33.7 million from
Kuwait

MINURSO
UN Mission for the Referendum in Western
Sahara
Since April 1991
Strength: military 229; international civilian
223; civilian police 34; local civilian 126
Fatalities: 10
Appropriation for July 2001–June 2002:
$50.5 million (gross)

UNOMIG
UN Oberserver Mission in Georgia
Since August 1993
Strength: military 106; international civilian
93; local civilian 169
Fatalities: 3
Appropriation for July 2001–June 2002:
$27.9 million (gross)

UNMIBH
UN Mission in Bosnia and Herzegovina
Since December 1995
Strength: military 4; international civilian
339; civilian police 1,680; local civilian
1,508
Fatalities: 8
Appropriation for July 2001–June 2002:
$144.7 million (gross); budget figures
include UNMIBH, UNMOP, and the UN
liaison offices in Belgrade and Zagreb

UNMOP
UN Mission of Observers in Prevlaka
Since January 1996
Strength: military 27; international civilian
3; local civilian 6
Fatalities: 0
Appropriation included in UNMIBH

UNMIK
UN Interim Administration Mission in Kosovo
Since June 1999
Strength: military 38; international civilian
1,228; civilian police 4,332; local civilian
3,176
Fatalities: 13
Appropriation for July 2001–June 2002:
$413.4 million (gross)

UNAMSIL
UN Mission in Sierra Leone
Since October 1999
Strength: military 16,654; international
civilian 297; civilian police 49; local
civilian 464
Fatalities: 46
Commitment authority for July
2001–December 2001 (six months):
$293.4 million (gross)[a]

(continues)

Table 5.1 continued

UNTAET
UN Transitional Authority in East
  Timor
Since October 1999
Strength: military 8,125; international
  civilian 972; civilian police 1,489; local
  civilian 1,859
Fatalities: 16
Commitment authority for July
  2001–December 2001 (six months):
  $300.8 million (gross)[a]

MONUC
UN Organization Mission in the Democratic
  Republic of Congo
Since December 1999
Authorized strength: military 5,537

Strength on August 31, 2001: military
  2,398; international civilian 475; local
  civilian 170
Fatalities: 2
Appropriation for July 2001–December
  2001 (six months): $209.1 million
  (gross)[a]

UNMEE
UN Mission in Ethiopia and Eritrea
Since July 2000
Strength: military 3,866; international
  civilian 235; local civilian 163
Fatalities: 2
Commitment authority for July
  2001–December 2001 (six months):
  $96.0 million (gross)[a]

Source: Adapted from www.un.org/depts/dpko/dpko/ops.htm.
Notes: UNTSO and UNMOGIP are funded from the United Nations regular budget.
Costs to the United Nations of the thirteen other current operations are financed from their
own separate accounts on the basis of legally binding assessments on all member states.
For these missions, budget figures are for one year (July 1, 2001–June 30, 2002) or for six
months (beginning on July 1, 2001) and include the prorated share of the support account for
peacekeeping operations and the United Nations Logisitics Base in Brindisi, Italy.
  a. Revised budget for the period July 1, 2001–June 30, 2002 in preparation.

and looting and a continuing struggle for power between factions—such as
in Somalia and in the former Yugoslavia—action by the UN, other groups,
the United States alone, or NATO has not been regarded as entirely successful. It has become increasingly clear that there must be a peace to keep
before peacekeeping can be useful.

Today's peacekeeping is described as "the power of persuasion backed
by force" (UNDPI 1998: 4). A UN mission may include military, civilian
police, and civilians. It may include such varied activities as truce supervision, cease-fire monitoring, military observation, disarmament, demobi-
lization and reintegration of former combatants, humanitarian assistance,
electoral assistance, education on human rights, mine clearance, and cooperation with regional organizations (UNDPI 1998: 15–16).

Following the tremendous rise in peacekeeping in the early 1990s, the
UN entered a period of reassessment and began to step back from its enormous investment and to draw conclusions on the lessons learned from
peacekeeping operations. Among these was that it is essential to have the
size and nature of the force fit the situation. The UN also began to realize
that peace could not be enforced from the outside and that there was no
substitute for political negotiations. In August 2000 the Special Panel on
UN Peace Operations produced a review and recommendations for reform

(commonly known as the "Brahimi Report"), and Secretary-General Kofi Annan followed with an implementation report (A/52/502), which will be considered by the Security Council and General Assembly.

■ CONDITIONS FOR SUCCESSFUL PEACEKEEPING

The chief conditions of success in peacekeeping, whether for restraining violence or for resolving conflict, are more likely to be political than military. A clear mandate adopted with the greatest degree of consensus possible and the consent of the parties to the conflict may be among the most important conditions. But there are other conditions that have developed out of the experience of UN peacekeeping.

The set of operating conditions that the UN Secretary-General attached to the deployment of the second United Nations Emergency Force (UNEF) in 1973 has come to be regarded as prerequisites for success. Deployment must be done only (1) with the full confidence and backing of the Security Council and (2) with the full cooperation and consent of the host countries. The force must (3) be under UN command through the Secretary-General; (4) have complete freedom of movement in the countries; (5) be international in composition, with national contingents acceptable to the parties in conflict; (6) act impartially; (7) use force only in self-defense; and (8) be supplied and administered under UN arrangements (UN 1973). Perhaps the most important conditions here are the nonuse of force except in self-defense and the consent of the parties to station a force. UNIFIL (UN Interim Force in Lebanon) and MONUC (UN Organization Mission in the Democratic Republic of Congo) have become classic examples of missions where several of the basic conditions were not met and where success was at best questionable.

It has been argued that it is not the use of force per se that is the problem but the unsuccessful use of force where the intervention does not receive the support of the majority of the population. A guideline for success may be that force can only be effectively used to restrain a very small minority of the population when that minority is violating agreed-upon norms, as is the case with the domestic use of police in participatory democratic societies.

Probably the most important factor in success, then, is not the mission on the ground but whether it is accompanied by sufficient efforts to facilitate the resolution of the underlying conflict that led to the violence. While participants in the actual peacekeeping operation may do harm, if they are unskilled, partial, badly commanded, or without sufficient resources or information, they cannot resolve the underlying conflict on their own.

Reducing violence and helping resolve conflict are both important criteria for success, but the cost factor must also be considered. Conflict is costly in terms of time, money, and opportunities forgone; violent conflict is even more costly, because it takes lives as well. Where conflicts remain unresolved, there is the constant potential for the resumption of violence. The cost of simply maintaining order without resolving underlying conflicts may not be worth it, either to the local or the international community. Perhaps a peacekeeping force, to provide real security, must always have both elements: the restraint of violence *and* assistance in the resolution of conflicts.

Other factors, such as leadership, organization, and communication, are also important. Adequate logistical support is critical. Secretary-General Boutros-Ghali (1995) made particular note of this in *An Agenda for Peace*. Adequate and timely financing is also crucial to the success of peacekeeping or any organizational mission. Where peacekeeping forces are in considerable arrears, as many are, this constrains their effectiveness both logistically and politically.

■ NONVIOLENT PEACEKEEPING

It is not certain that military personnel are the best peacekeepers. Brigadier General Michael Harbottle made the point that one of the lessons of UNEF, MONUC, and UNFICYP (UN Peacekeeping Force in Cyprus) was that the professional soldier was no better as a peacekeeper than a volunteer. Volunteers are more likely to want to be there, and because they come from all walks of life, they may be more likely to have a common bond with those in the communities in which they are deployed (Harbottle 1971). Among the factors critical to the success of an operation are the attitudes of the members of the force, their ability to be sensitive to cultural differences, and their ability to solve problems and to facilitate the resolution of conflicts. Increasingly, the composition of peacekeeping missions is more and more diverse, including not only military personnel and civilian police, but administrators, observers, and other civilian personnel.

Civilians trained in the nonviolent resolution of conflict may have an increasingly important role to play in international peacekeeping, peacemaking, and peace-building, especially in the context of community-wide conflict. Nongovernmental nonviolent peacekeeping forces organized on the Gandhian *shanti sena* (peace army) model have sent volunteers into situations of violence to monitor human rights or border crossings (see Desai 1972). Peace Brigades International is one of these, with an international board from all continents. Originally called the World Peace Brigade, it was founded in Beirut, Lebanon, on January 1, 1962, and has sent transnational teams of observers to conflicts in Africa, Southeast Asia, and Central America. The conflict between the goals of reconciliation through mediation (which stresses impartiality) and confrontation in the name of justice (which tends to stress partisan support of one side of the struggle) has been an issue of debate within Peace Brigades International.

More and more unarmed nongovernmental peacekeeping groups have begun to play roles in conflict areas. Witness for Peace has operated both as a nonviolent witness and as a tripwire and communications device to keep U.S. citizens aware of what the United States and other governments are doing or planning in Central America. It has sent observers as support for local communities that are working for development and human rights in the face of repressive governments. The Cyprus Resettlement Project has worked on reconciliation and development, the Balkan Peace Team has worked in Croatia, Cry for Justice in Haiti, and Christian Peacemaker Teams in a variety of settings. As nongovernmental organizations become increasingly involved in day-to-day policymaking and administration in the United Nations, perhaps there is a role that trained, unarmed civilians skilled in nonviolent action and conflict resolution can play to improve the UN's peacemaking and peacekeeping capacity on the ground.

■ CONCLUSION

Perhaps the dilemmas the UN faces would not be so difficult if the international system were not so biased toward the utility of violent force. The efficacy of violence is a myth: at best, only some win—and at the great expense of others. Because violent force does not make peace or justice, its use by the UN would not seem to be the most cost-effective use of its limited resources.

It may well be that the situation the UN is in with respect to funding and general support may, in contrast to popular thinking, be due to the overuse and misuse of violent force rather than to its inability to bring about international (and subnational) peace and security. The Nobel Peace Prize was awarded after a period of many small, less violent peacekeeping missions. Criticism has come following the UN missions in Somalia and the former Yugoslavia. Perhaps it is time to have a more pointed debate on the merits of long- versus short-term, and more violent versus less violent, approaches to international peace, security, and justice.

If the international system is ready to move in the direction of international peace, security, and justice, it must overcome the myth of the efficacy of violence and take on the more difficult task of creating security cooperatively. Less violent, longer-term means for providing international peace and security, such as peacemaking and peace-building, may hold much more promise than traditional enforcement—or even peacekeeping.

■ DISCUSSION QUESTIONS

1. How have the concepts of peace and security evolved?
2. Do you think peacekeeping and peacemaking will be more attainable now that the Cold War is over?
3. Is it in the interest of the United States to cooperate with the United Nations in its peacekeeping and peacemaking efforts?
4. Should the United Nations have its own military force? How might this raise concerns over state sovereignty?
5. Could nonviolent approaches be useful as a means of bringing about peace?

■ SUGGESTED READINGS

Boutros-Ghali, Boutros (1995) *An Agenda for Peace*. Second edition. New York: United Nations.

Harbottle, Michael (1971) *The Blue Berets*. London: Leo Cooper.

Higgins, Rosalyn (1996) *United Nations Peacekeeping*. 4 vols. London: Oxford University Press.

International Peace Academy (1984) *Peacekeeper's Handbook*. New York: Pergamon Press.

Mahony, Liam, and Luis Enrique Eguren (1997) *Unarmed Bodyguards*. West Hartford, CT: Kumarian Press.

Moser-Puangsuwan, Yeshua, and Thomas Weber (2000) *Nonviolent Intervention Across Borders*. Honolulu: University of Hawaii Press.

Ramsbottom, Oliver, and Tom Woodhouse (1999) *Encyclopedia of International Peacekeeping Operations*. Santa Barbara, CA: ABC-CLIO, Inc.

UN Department of Public Information (UNDPI) (1996) *The Blue Helmets: A Review of United Nations Peace-Keeping*. Third edition. New York: UNDPI.

———— (1998) *Peacekeeping: 50 Years 1948–1998*. New York: UNDPI (Pamphlet DPI/2004, 88 pages).

UN Panel on UN Peace Operations (August 21, 2000) *Report of the Panel on UN Peace Operations*. UN Resolutions A/55/305-S/2000/809 (the "Brahimi Report").

Weiss, Thomas G., ed. (1993) *Collective Security in a Changing World*. Boulder, CO: Lynne Rienner Publishers.

White, N. D. (1990) *The United Nations and the Maintenance of International Peace and Security*. New York: Manchester University Press.

Wiseman, Henry (1983) "United Nations Peacekeeping: An Historical Overview." In Henry Wiseman, ed., *Peacekeeping, Appraisals and Proposals*. New York: Pergamon Press.

Part 2

THE GLOBAL ECONOMY

6

CONTROVERSIES IN INTERNATIONAL TRADE

Bruce E. Moon

International trade is often treated purely as an economic matter that can and should be divorced from politics. That is a mistake, because trade not only shapes our economy but also determines the kind of world in which we live. The far-reaching consequences of trade pose fundamental choices for all of us. Certainly citizens must understand those consequences before judging the inherently controversial issues that arise over trade policy. More than that, we cannot even make sound consumer decisions without weighing carefully the consequences of our own behavior.

■ THE CASE FOR TRADE

The individual motives that generate international trade are familiar. Consumers seek to buy foreign products that are better or cheaper than domestic ones in order to improve their material standard of living. Producers sell their products abroad to increase their profit and wealth.

Most policymakers believe that governments should also welcome trade because it provides benefits for the nation and the global economy as well as for the individual. Exports produce jobs for workers, profits for corporations, and revenues that can be used to purchase imports. Imports increase the welfare (well-being) of citizens because they can acquire more for their money as well as obtain products that are not available from domestic sources. The stronger economy that follows can fuel increasing power and prestige for the nation as a whole. Further, the resultant interdependence and shared prosperity among countries may strengthen global cooperation and maintain international peace.

Considerable evidence supports the view that trade improves productivity, consumption, and therefore material standard of living (Moon 1998). Trade successes have generated spurts of national growth, most notably in East Asia. The global economy has grown most rapidly during periods of trade expansion, especially after World War II, and has slowed when trade levels have fallen, especially during the Great Depression of the 1930s. Periods of international peace have also coincided with trade-induced growth, while war has followed declines in trade and prosperity. More recent evidence suggests that trade expansion may be considerably more beneficial for developed nations than for poorer ones, however. Indeed, a 1999 World Bank study reported that the median per capita growth of developing countries was 0.0 percent since 1980, that is, during the era of globalization in which trade levels exploded (Easterly 1999).

Still, the private benefits of trade have led individual consumers and producers to embrace it with zeal for the last half-century. As a result, trade has assumed a much greater role in almost all nations, with exports now constituting about a quarter of the economy in most countries and over half in the developing world (World Bank 1999: 222). Even in the United States, which is less reliant on trade than virtually any other economy in the world because of its size and diversity, the export sector is now about 10 percent of gross domestic product (GDP), defined as the total of goods and services produced in a given year. Smaller nations must engage in more trade because they can neither supply all their own needs nor provide a market sizable enough for their own industries to operate efficiently.

Since World War II, most governments have encouraged and promoted this growth in trade levels. All but a handful of nations now rely so heavily on jobs in the export sector and on foreign products to meet domestic needs that discontinuing trade is no longer an option. To attempt it would require a vast restructuring that would entail huge economic losses and massive social change. Furthermore, according to the "liberal" trade theory accepted by most economists, governments have no compelling reason to interfere with the private markets that achieve such benefits. The reader is cautioned that *liberalism*, as used throughout this chapter, refers to liberal economic theory that opposes government interference with the market and is not to be confused with the ambiguous way the term *liberal* is applied in U.S. politics, where it often means the opposite.

From its roots in the work of Scottish political economist Adam Smith (1723–1790) and English economist David Ricardo (1772–1823), this liberal perspective has emphasized that international trade can benefit all nations simultaneously, without requiring governmental involvement (Smith 1910). According to Ricardo's theory of *comparative advantage* (Ricardo 1981), no nation need lose in order for another to win, because trade allows total global production to rise. The key to creating these gains from trade is the efficient allocation of resources, whereby each nation specializes in the production of goods in which it has a comparative advantage. For example, a nation with especially fertile farmland and a favorable climate can produce food much more cheaply than a country that lacks this comparative advantage. If it were to trade its excess food production to a nation with efficient manufacturing facilities for clothing production, both nations would be better off, because trade allows each to apply its resources to their most efficient use. No action by governments is required to bring about this trade, however, since profit-motivated investors will see to it that producers specialize in the goods in which they have a comparative advantage, and consumers will naturally purchase the best or cheapest products. Thus, liberal theory concludes that international trade conducted by private actors free of government control will maximize global welfare.

■ CHALLENGES TO THE LIBERAL FAITH IN TRADE

Though trade levels have grown massively in the two centuries since Adam Smith, no government has followed the advice of liberal economic theorists to refrain from interfering with trade altogether. That is because governments also have been influenced by a dissenting body of thought known as *mercantilism,* which originated with the trade policy of European nations, especially England, from the sixteenth century to the middle of the nineteenth.

While mercantilists do not oppose trade, they *do* hold that governments must regulate it in order for trade to advance various aspects of the national interest. The aspirations of mercantilists go beyond the immediate consumption gains emphasized by liberals to include long-term growth, national self-sufficiency, the vitality of key industries, and a powerful state in foreign policy. Because most states accept the mercantilist conviction that trade has negative as well as positive consequences, they try to manage it in a fashion that will minimize its most severe costs yet also capture the benefits claimed for it by liberal theory.

In particular, mercantilists observe that the rosy evaluation of trade advanced by Smith and Ricardo was predicated on their expectation that any given nation's imports would more or less balance its exports. However, when a nation's imports are *greater* than its exports—meaning that its residents buy more from other nations than it sells to them—mercantilists warn that this "trade deficit" carries with it potential dangers that may not be readily apparent. On its face a trade deficit appears as the proverbial free lunch: If a nation's imports are greater than its exports, it follows that national consumption must exceed its production. One might ask how anyone could object to an arrangement that allows a nation to consume more than it produces. The answer lies in recognizing that such a situation has adverse repercussions (especially in the future), just as individuals cannot continue to spend more than they earn without eventually suffering detrimental consequences.

For example, the United States has run a substantial trade deficit for three decades, with imports surpassing exports by about $1 billion *per day* in 2000, a trade deficit that annually allows U.S. citizens to enjoy a standard of living more than $1,000 per person higher than would otherwise be possible. But mercantilists fear that excessive imports permit foreigners to enjoy employment and profits from production that might otherwise benefit citizens of the home country. For example, since the U.S. trade deficit began to bloom in the 1970s, the massive sales of Japanese cars in the United States have transferred millions of jobs out of the U.S. economy, accounting for high levels of unemployment in Detroit and low levels of unemployment in Tokyo.

However, the greater anxiety over trade deficits concerns their longer-term impact, because they generate future liabilities just as any other form of indebtedness. Unfortunately, the consequences of trade imbalances cannot be evaluated easily because they trigger complex flows of money that are also unbalanced and unpredictable. To understand this point, consider that the trade deficit of the United States means that more money flows *out* of the U.S. economy in the form of dollars to pay for imports than flows back *into* the U.S. economy through payments for U.S. goods purchased by foreigners. The consequences of the trade deficit depend in large part on what happens to those excess dollars, which would appear to be piling up abroad. Eventually, one assumes, foreigners will want to use those dollars to purchase U.S. goods, and when they do, the result could be catastrophic. Suddenly, Americans who have grown accustomed to consuming far more than they produce will be forced to consume far less.

In practice, many of those dollars have already found their way back into the U.S. economy, because foreigners have used them to purchase U.S. Treasury bonds and real estate and to finance the takeover of U.S. businesses as well as to make new investments in the United States. Such capital flows can offset a trade deficit temporarily and render it harmless in the short run, but they create future liabilities that only postpone the inevitable need to balance production and consumption. For example, during the 1980s and 1990s, the U.S. Treasury borrowed hundreds of billions of dollars from foreigners by selling them U.S. Treasury bonds, a form of debt that must someday be repaid. In the meantime, foreigners now receive about $70 billion in interest payments annually from the U.S. federal government (Forsyth 1996). Moreover, the other investments made by foreigners out of the proceeds of their past trade surpluses with the United States also represent future liabilities, even though they do not constitute "debt" in the same formal accounting sense. For example, Honda auto plants located in the United States can be expected to return millions of dollars to their owners in Japan every year far into the future. This constant outflow of dollars resulting from past trade deficits only adds to the outflow associated with the current one. Overall, the net liability of U.S. public and private debt to foreigners now amounts to well over $2 trillion, and it continues to mount at a rate equivalent to the annual trade deficit, more than $300 billion per year recently.

Economists disagree about whether these developments ought to raise alarm. After all, the willingness of foreigners to invest in the United States surely is an indication of confidence in the strength of the U.S. economy. More generally, as Chapter 7 shows, capital flows can be beneficial to the economy and its future. Indeed, foreign capital is an essential ingredient to development in many third world countries. Whether capital inflows produce effects that are, on balance, positive or negative depends heavily on the source of the capital, the terms on which it is acquired, and especially on the uses to which it is put.

In the case of the United States, international currency markets have sent a cautionary signal that the persistent trade deficit is eroding the confidence of foreign investors. Even with massive inflows of investment from abroad, the demand for dollars by foreigners to purchase products or investments from the United States has been smaller than the supply of dollars created by Americans purchasing foreign products and investments. As a result, the value of the dollar, once equivalent to 360 Japanese yen (¥360), declined to under ¥80 in early 1995 (before recovering to ¥125 in early 2002). These fluctuations in the purchasing power of the dollar mean, for example, that the ¥2,000,000 cost of a Japanese automobile would translate into a dollar price of about U.S.$5,600 at the exchange rate of ¥360 per dollar, but would require more than U.S.$20,000 when the dollar fell below ¥100.

Balance-of-trade deficits tend to lead to such currency declines and to both future price increases and snowballing debt. Thus, a trade deficit provides immediate benefits but also implies that future consumption may be reduced so that the standard of living for future generations will fall. Because these consequences are uncertain, nations vary somewhat in their tolerance for trade deficits, but most try to minimize or avoid them altogether, as counseled by mercantilists.

■ OPTIONS IN TRADE POLICY

To achieve their desired trade balance, nations often combine two mercantilist approaches. They may emphasize the expansion of exports through a strategy known as *industrial policy*. More commonly, they emphasize minimizing imports, a stance known generally as *protectionism* (Fallows 1993).

Protectionist policies include many forms of import restriction designed to limit the purchase of goods from abroad. All allow domestic import-competing industries to capture a larger share of the market and, in the process, to earn higher profits and to employ more workers at higher wages. The simplest import barriers are *quotas*, government restrictions that place a fixed limit on the quantity or value of goods that can be imported. The most traditional barriers are taxes on imports called *tariffs* or import duties, but they are no longer the main form of protectionism in most countries.

In fact, declining from their peak in the 1930s, tariff levels throughout the world are now generally very low. In the United States, the average tariff rate reached a modern high of 59 percent in 1932 under what has been called "a remarkably irresponsible tariff law," the Smoot-Hawley Act, which has been widely credited with triggering a spiral of restrictions by other nations that helped plunge the global economy into the Great Depression of the 1930s. The average rate in the United States was reduced to 25 percent after World War II and declined to about 2 percent after the Uruguay Round of trade negotiations (discussed in greater detail later in the chapter) concluded in 1994. Most other countries have followed suit—and some have reduced even more—so that average rates above 10 percent are now very rare.

However, in place of tariffs, governments have responded to the pleas of industries threatened by foreign competition with a variety of *nontariff barriers* (NTBs), especially *voluntary export restraints* (VERs). In the most famous case of VERs, Japanese automakers "voluntarily" agreed to limit exports to the United States in 1981. (Had Japan refused, a quota that would have been more damaging to Japanese automakers would have been imposed.) The Federal Trade Commission (FTC) has estimated that the higher prices for autos that resulted cost U.S. consumers about $1 billion per year. Not only does the restricted supply of Japanese autos cause their prices to rise because of the artificial shortage, but it also enables U.S. manufacturers to maintain higher prices in the absence of this competition.

A favorable trade balance (or the elimination of an unfavorable one) also can be sought through an industrial policy that promotes exports. The simplest technique is a *direct export subsidy*, in which the government pays a domestic firm for each good exported, so that it can compete with foreign firms that otherwise would have a cost advantage. Such a policy has at least three motivations. First, by increasing production in the chosen industry, it reduces the unemployment rate. Second, by enabling firms to gain a greater share of foreign markets, it gives them greater leverage to increase prices (and profits) in the future. Third, increasing exports will improve the balance of trade and avoid the problems of trade deficits.

Liberals are by no means indifferent to the dangers of trade deficits, but they argue that most mercantilist cures are worse than the disease.

When mercantilist policies affect prices, they automatically create winners and losers and in the process engender political controversies. For example, to raise the revenue to pay for a subsidy, the domestic consumer has to pay higher taxes. As noted above, protectionism also harms the consumer by raising prices even while it benefits domestic firms that compete against imports.

If mercantilist policies are controversial in the nations that enact them, they are met with even greater hostility by the nations with which they trade. For example, U.S. firms complain that Japanese protectionism prevents them from competing for the lucrative Japanese market, while the export promotion policies of many countries place U.S. firms at a disadvantage even in the U.S. market. The U.S. steel industry has been particularly outspoken in its denunciation of steel imported from foreign firms, especially those that are heavily subsidized by their governments. They contend that U.S. jobs and U.S. profits are being undercut unfairly and, in December 2001, the U.S. International Trade Commission recommended a tariff to protect the U.S. steel industry from foreign competition.

The United States has generally preferred to maintain a desirable volume and balance of trade by inducing other nations to lower their trade barriers rather than by erecting its own. The United States has undertaken direct bilateral negotiations to change the policies of other nations, especially Japan, and has spearheaded efforts to create and maintain global institutions that facilitate trade, as discussed below.

■ THE MULTIPLE CONSEQUENCES OF TRADE

As nations choose among policy options, they must acknowledge liberal theory's contention that free trade allows the market to efficiently allocate resources and thus to maximize global and national consumption. Neither can the desire of individual consumers and producers to participate in trade be ignored. As my brief survey of the consequences of a trade deficit illustrates, however, the simplicity of individual motives conceals the complexity of the effects that trade has on others. As I will illustrate, the dangers of trade deficits are only a small part of why governments almost universally restrict trade, at least to some degree.

In fact, governments seek many outcomes from trade—full employment, long-term growth, economic stability, social harmony, power, security, and friendly foreign relations—yet discover that these desirable outcomes are frequently incompatible with one another. Because free trade may achieve some goals but undermine others, governments that fail to heed the advice of economic theory need not be judged ignorant or corrupt. Instead, they recognize a governmental responsibility to cope with all of trade's consequences, not only those addressed by liberal trade theory. For example, while trade affects the prices of individual products, global markets also influence which individuals and nations accumulate wealth and political power. Trade determines who will be employed and at what wage. It determines what natural resources will be used and at what environmental cost. It shapes opportunities and constraints in foreign policy.

Because trade affects such a broad range of social outcomes, conflict among alternative goals and values is inevitable. As a result, both individuals and governments must face dilemmas that involve the multiple consequences of trade, the multiple goals of national policy, and the multiple values that compete for dominance in shaping behavior (Moon 2000).

■ The Distributional Effects of Trade, or Who Wins, Who Loses?

Many of these dilemmas stem from the sizable effect that international trade has on the distribution of income and wealth among individuals, groups, and nations. Simply put, some gain material benefits from trade while others lose. Thus, to choose one trade policy and reject others is simultaneously a choice of one income distribution over another. As a result, trade is inevitably politicized: Each group pressures its government to adopt a trade policy from which it expects to benefit.

The most visible distributional effects occur because trade policy often protects or promotes one *industry or sector* of the economy at the expense of others. For example, tariffs on imported steel protect the domestic steel industry by making foreign-produced steel more expensive, but they also harm domestic automakers who must pay higher prices for the steel they use. As in this case—where car buyers face higher prices as a result—most barriers to trade benefit some sector of the economy at the expense of consumers, a point always emphasized by proponents of free trade.

Trade policy also benefits some *classes and regions* at the expense of others, a point more often emphasized by those who favor greater governmental control. For example, the elimination of trade barriers between the United States and Mexico under the terms of the North American Free Trade Agreement (NAFTA) forces some U.S. manufacturing workers into direct competition with Mexican workers, who earn a markedly lower wage. Since NAFTA guarantees that imports can enter the United States without tariffs, some U.S. businesses move to Mexico where production costs are lower, and U.S. workers lose their jobs in the process. Facing the threat of such production shifts, many more U.S. workers will accept a decline in wages, benefits, or working conditions. The losses from such wage competition will be greatest for unskilled workers in high-wage countries employed in industries that can move either their products or their production facilities most easily across national boundaries. Others, particularly more affluent professionals who face less direct competition from abroad (such as doctors, lawyers, and university professors), stand to gain from trade because it lowers prices on the goods they consume. Of course, the greatest beneficiaries are the owners of businesses that profit from lower wage rates and expanded markets.

Proponents of free trade tend to de-emphasize these distributional effects and instead focus on the impact of trade on the economy as a whole. That is partly because liberal theory contends that free trade does not decrease employment but only shifts it from an inefficient sector to one in which a nation has a comparative advantage. For example, U.S. workers losing their jobs to Mexican imports should eventually find employment in industries that export to Mexico. Proponents of free trade insist that it is far better to tolerate these "transition costs"—the short-term dislocations and distributional effects—than to protect an inefficient industry.

Because these distributional consequences have such obvious political implications, however, the state is much more attentive to them than economic theorists are. That is one reason all governments control trade to one degree or another. Of course, that does not mean that they do so wisely or fairly, in part because their decisions are shaped by patterns of representation among the constituencies whose material interests are affected by trade policy. In general, workers tend to be underrepresented, which is why trade policies so often encourage trade built on low wages that enrich business owners but constrain the opportunities for workers. Similarly, the economic structures created by trade patterns can produce just as great a distributional inequity between genders as between classes, sectors, or regions. As Chapter 10 describes, political representation of women in decisionmaking is very poor. Finally, as the discussion of trade deficits indicated, the economic activities shaped by trade policies tend to affect current generations very differently from future ones—and the latter are seldom represented at all.

■ *The Values Dilemma*

These distributional effects pose challenging trade-offs among competing values. For example, the effects of NAFTA were predicted to include somewhat lower prices for U.S. consumers but also job loss or wage reduction for some unskilled U.S. workers. The positions taken on this issue by most individuals, however, did not hinge on their own material interests; few could confidently foresee any personal impact of NAFTA since the gains were estimated at well under 1 percent of GDP, and job losses were not expected to exceed a few hundred thousand in a labor force of more than 100 million. However, the choice among competing *values* was plain: NAFTA meant gains in wealth but also greater inequality and insecurity for workers. Some citizens acceded to the judgment of liberal theory that the country as a whole would be better off with freer trade, while others identified with the plight of workers, who were more skeptical of liberal theory simply because for them the stakes were so much higher. After all, it is easy for a theorist to postulate that job losses in an import-competing industry would be matched by job gains in an exporting firm but far harder for a worker who has devoted his life to one career to pack up and move to a strange town, hopeful that he *might* find a job that requires skills he may not possess in an unfamiliar industry. In the final analysis, NAFTA became a referendum on what kind of society people wished to live in. The decision was quintessentially American: one of greater wealth but also greater inequality and insecurity.

Of course, other distributional effects gave rise to other value choices as well. Since the gains from NAFTA were expected to be greater for Mexico than for the United States, the conscientious citizen would also weigh whether it is better to help Mexican workers because they are poorer or to protect U.S. workers because they are U.S. citizens. As Chapter 8 implies, such issues of inequality in poor societies can translate directly into questions of life or death. As a result, the importance of trade policy, which has such a powerful impact on the distribution of gains and losses, is heightened in poor, dependent nations where half of the economy is related to trade.

Perhaps the most challenging value trade-offs concern the trade policies that shift gains and losses from *one time period to another.* Such "intergenerational" effects arise from a variety of trade issues. For example, as discussed earlier, the U.S. trade deficit, like any form of debt, represents an immediate increase in consumption but a postponement of its costs. The Japanese industrial policy of export promotion fosters a trade surplus, which produces the opposite effect in Japan. The subsidies the Japanese government pays to Japanese exporters require Japanese citizens to pay both higher prices and higher taxes. However, the sacrifices of Japan's current generation may benefit future ones if this subsidy eventually transforms an "infant industry" into a powerful enterprise that can repay the subsidies through cheaper prices or greater employment. Interestingly, those subsidies make Japanese products cheaper abroad, so consumers in countries like the United States benefit, reinforcing the U.S. preference for immediate gratification. Of course, if the subsidies drive U.S. firms out of business, future generations of U.S. citizens may suffer losses of employment opportunities and higher prices. It is interesting to speculate why U.S. policies so frequently differ from Japanese ones when distributional effects pose the dilemma of whether it is better to sacrifice now for the future or to leave future generations to solve their own problems.

The values dilemma encompasses much more than just an alternative angle on distributional effects, however (Polanyi 1944). The debate over "competitiveness," which began with the efforts by U.S. businesses to lower their production costs in order to compete with foreign firms, illustrates how trade may imply a compromise of other societal values. Companies often find that government policies make it difficult for them to lower their labor costs. Lower wages could be paid if the minimum wage were eliminated and collective bargaining and labor unions were outlawed. The abolition of seniority systems and age discrimination laws would enable companies to terminate workers when their efficiency declined (or at the whim of a boss). Eliminating pensions, health care, sick leave, workers' compensation for accidents, workplace safety regulations, and paid vacations and holidays would also lower company labor costs. But such actions entail a compromise with fundamental values about the kind of society in which people want to live. Government regulations that handicap business were designed to meet other legitimate national goals, addressing such issues as equality, security, social harmony, and ecological sustainability. Environmental regulations, for example, may add to production costs, but surely achieving economic interests is not worth abandoning all other values. Choosing between alternative values is always difficult for a society, because reasonable people can differ in the priority they ascribe to each. Still, agreements on such matters can usually be forged *within* societies, in part because values tend to be broadly, if not universally, shared.

Unfortunately, free trade forces firms burdened by these value choices in one country to compete with firms operating in countries that may *not* share them. This situation creates a dilemma for consumers, forcing them to balance economic interests against other values. For example, continuing to trade with nations that permit shabby treatment of workers—or even outright human rights abuses—poses a painful moral choice, not least because goods from such countries are often cheaper. As Chapter 4 documents, foreign governments have often declared their opposition to human rights abuses but have seldom supported their rhetoric with actions that effectively curtailed the practice. In fact, the implementation of the policy to pressure foreign governments on behalf of a normative stance has been left to consumers, who have unwittingly answered key questions daily: Should we purchase cheap foreign goods like clothing and textiles even though they may have been made with child labor—or even slave labor? Of course, we seldom know the conditions under which these products were produced—or even *where* they were produced—so we pressure our government to adopt policies on behalf of principles we cannot personally defend.

Where values are concerned, of course, we cannot expect everyone to agree with the choices we might make. As Chapter 11 describes, child labor remains a key source of comparative advantage for many countries in several industries prominent in international trade. We cannot expect them to give up easily a practice that is a major component of their domestic economy and that is more offensive to us than to them. Unfortunately, if trade competitors do not share our values, it may prove difficult to maintain them ourselves—unless we restrict trade, accept trade deficits, or design state policies to alleviate the most dire consequences. After all, it is hard to see how U.S. textile producers can compete with the sweatshops of Asia without creating sweatshops in New York. That point inevitably animates a complex debate over whether eliminating sweatshops would really benefit the poor, a dilemma of international trade that cannot be avoided merely by refusing to think about it.

■ Foreign Policy Considerations: Power and Peace

Some of the most challenging value choices concern the effect of trade on the foreign policy goals pursued by states, especially power, peace, and national autonomy. For example, policymakers have long been aware that trade has two deep, if contradictory, effects on national security. On the one hand, trade contributes to national prosperity, which increases national power and enhances security. On the other hand, it has the same effect on a nation's trade partner, which could become a political or even military rival. The resulting ambivalent attitude is torn between the vision of states cooperating for economic gain and the recognition that they also use trade to compete for political power.

While a market perspective sees neighboring nations as potential customers, the state must also see them as potential enemies. As a result, the state not only must consider the absolute gains it receives from trade but also must weigh those gains in relative terms, perhaps even avoiding trade that would be more advantageous to its potential enemies than to itself. For this reason, states have always been attentive to the distribution of the gains from trade and selective about their trade partners, frequently encouraging trade with some nations and discouraging or even banning it with others.

While understandable, such policies create competitive struggles for markets, raw materials, and investment outlets, which sometimes can lead to open conflict. In fact, U.S. president Franklin Roosevelt's secretary of state Cordell Hull even went so far as to contend that bitter trade rivalries were the *chief* cause of World War I and a substantial contributor to the outbreak of World War II. Both were precipitated by discriminatory trade policies in which different quotas or duties were imposed on the products of different nations. Hull, who believed that free multilateral trade would build bridges rather than create chasms between peoples and nations, thus championed the nondiscrimination principle and urged the creation of international institutions that would govern international trade in accordance with it.

In fact, international institutions are absolutely essential to maintaining *any* kind of international trading system, let alone one that reduces the conflict potential among nations. The need for such institutions at the global level is illustrated by the Bretton Woods trade and monetary regime created under the leadership of the United States at the end of World War II. It is centered on the institutions of the General Agreement on Tariffs and Trade (GATT), the World Bank, and the International Monetary Fund (IMF). Institutions are necessary to overcome the inclination of most nations to retain their own trade barriers while inducing other countries to lower theirs. Since 1946, the GATT has convened eight major negotiating sessions (referred to as rounds) in which nations exchange reductions in trade barriers, with the result that global trade has increased dramatically.

Institutions are also necessary to provide a stable monetary system that facilitates trade by permitting the easy exchange of national currencies and the adjustment of trade imbalances, a role taken up at the global level by the IMF. The World Bank lends money to nations that might otherwise seek trade-limiting solutions to their financial problems. At the regional level the European Union (EU) has struggled with monetary issues for years, finally fashioning a new regional currency, the euro, to replace national currencies beginning in 2001. Finally, international institutions are necessary to establish the rules of trade, create the international law that embodies them, and provide a forum for resolving the disputes that inevitably arise. Thus, the Uruguay Round of GATT, completed in 1994, created the World Trade Organization (WTO) to provide a global setting to resolve trade disputes among its 143 members peacefully.

Regionally, a similar belief in the efficacy of free trade as a guarantor of peace was an important motivation for the initiative that eventually led to the European Union. This process was launched in 1951 with the founding of the European Coal and Steel Community (ECSC), which internationalized an industry that was key not only for the economies of the six nations involved but also for their war-making potential. With production facilities scattered among different countries, each became dependent on the others to provide both demand for the final product and part of the supply capacity. This arrangement fulfilled the liberal dream of an interdependence that would prevent war by making it economically suicidal. In fact, the ECSC was an innovative form of peace treaty, designed, in the words of Robert Schuman, to "make it plain that any war between France and Germany becomes, not merely unthinkable, but materially impossible" (Pomfret 1988: 75).

In both the EU and Bretton Woods, policymakers saw several ways that an institutionalized liberal trading system could promote peace among nations. The growth of global institutions could weaken the hold of nationalism and mediate conflict between nations. Trade-induced contact could break down nationalistic hostility among societies. Multilateralism (nondiscrimination) would tend to prevent grievances from developing among states. Interdependence could constrain armed conflict, and foster stability, while the economic growth generated by trade could remove the desperation that leads nations to aggression.

Such peaceful pursuits require the exercise of power. According to hegemonic stability theory, one dominant nation—a hegemon—will usually have to subsidize the organizational costs and frequently offer side benefits in exchange for cooperation, such as the massive infusion of foreign aid provided to Europe by the United States under the Marshall Plan in the late 1940s. Maintaining the capability to handle these leadership requirements entails substantial costs. For example, U.S. expenditures for defense, which have been many times higher than those of nations with whom it competes since World War II, erode the competitiveness of U.S. business by requiring higher tax levels; they constrain the funds available to spend on other items that could enhance competitiveness; and they divert a substantial share of U.S. scientific and technological expertise into military innovation and away from commercial areas. (Ironically, much of that money has been spent directly on protecting the very nations against whom U.S. competitiveness has slipped, especially Germany, Japan, and Korea.) The trade-off between competitiveness and defense may be judged differently by different individuals, but it can be ignored by none.

Neither can we neglect the complex relationship between peace and power raised by trade issues. The United States certainly sacrifices some power to maintain these organizations, but it also achieves considerable benefits from its capacity to strongly influence—and sometimes dictate—the rules under which they operate. Indeed, many of the critics who have protested at recent meetings of these institutions see them as extensions of U.S. imperialism.

◼ International Cooperation and National Autonomy

International institutions may be necessary to facilitate trade and to alleviate the conflict that inevitably surrounds it, but they also create conflict of their own. At issue is the tension between maintaining fair competition among firms in different countries—which is essential to sustaining the international trading system—and maintaining the cultural and political differences among nations—which is central to the national sovereignty and autonomy of the modern state system.

Fair competition in trade requires at least implicit cooperation between governments, because no nation can export unless some other nation imports. However, while nations usually encourage the exports on which they rely for jobs, for profits, and for the limitation of balance-of-trade deficits, they are usually less enthusiastic about welcoming imports. Fortunately, almost all nations acknowledge, in principle, the obligation to permit the sale of foreign products within their borders, if only because they fear that excessive protectionism of their own market will encourage other nations to protect theirs.

Still, disputes over trade barriers are common, because, in practice, governments have many compelling motives for enacting policies that affect trade. Often one nation defends its policy as a rightful exercise of national sovereignty, while another challenges it as an unfair barrier to trade. Ideally, such disagreements have been settled by appeal to the GATT or, more recently, to the WTO, whose new dispute resolution panel hears trade disputes and determines whether national behavior is consistent with international rules. But not even the WTO's chief sponsor, the United States, accepts the dominion of the WTO without serious reservations about its intrusion into affairs historically reserved for national governments. Since trade touches so many other areas of governmental responsibility, it seems unavoidable that this conflict would arise from time to time.

While the U.S. administration strongly supported the creation of the WTO to prevent trade violations by other nations, a surprising variety of U.S. groups opposed its ratification because it might encroach on national sovereignty. Environmental groups such as Friends of the Earth, Greenpeace, and the Sierra Club were joined not only by consumer advocates like Ralph Nader but also by conservatives such as Ross Perot, Pat Buchanan, and Jesse Helms, who feared that a WTO panel (a small panel of judges who hear disputes) could rule that various U.S. government policies constituted unfair trade practices, even though they were designed to pursue values utterly unrelated to trade. For example, EU automakers have challenged the U.S. law that establishes standards for auto emissions and fuel economy. Buchanan said, "WTO means putting America's trade under foreign bureaucrats who will meet in secret to demand changes in United States laws. . . . WTO tramples all over American sovereignty and states' rights" (Dodge 1994). Because the WTO could not force a change in U.S. law, GATT director-general Peter Sutherland called this position "errant nonsense" (Tumulty 1994), but the WTO could impose sanctions or authorize an offended nation to withdraw trade concessions as compensation for the injury.

The most dramatic example occurred in the 1994 case known as "GATTzilla versus Flipper," in which a GATT tribunal ruled in favor of a complaint brought by the EU on behalf of European tuna processors who buy tuna from Mexico and other countries that use purse seine nets. The United States boycotts tuna caught in that way because the procedure also kills large numbers of dolphins; but this value is not universally shared by other nations. In fact, the GATT ruled that the U.S. law was an illegal barrier to trade because it discriminates against the fishing fleets of nations that use this technique. The United States saw this as an unwarranted intrusion into its domestic affairs and an affront to U.S. values.

Soon thereafter the United States found itself on the other side of the clash between fair competition and national sovereignty when it appealed

to the WTO to rule that the EU's prohibition of beef containing growth hormones violated the "national treatment" principle contained in GATT's Article 3. Since almost all cattle raised in the United States are fed growth
hormones and very few European cattle are, the United States contended
that the EU rule was simply disguised protectionism that unfairly discrimi-
nated against U.S. products. The EU contended that such beef was a cancer risk and that as a sovereign power it had the right to establish whatever
health regulations it chose to protect its citizens. The WTO ruled in favor of
the United States, incurring the wrath of those who saw this as an example
of national democratic processes being overruled by undemocratic global
ones. Can it be long before Colombia challenges U.S. drug laws as dis-
criminating against marijuana while favoring Canadian whiskey?

Neither can regional agreements avoid this clash between fair competition in trade and national autonomy. The first trade dispute under NAFTA
involved a challenge by the United States to regulations under Canada's
Fisheries Act established to promote conservation of herring and salmon
stocks in Canada's Pacific Coast waters. Soon thereafter the Canadian government challenged U.S. Environmental Protection Agency (EPA) regulations that require the phasing out of asbestos, a carcinogen no longer permitted as a building material in the United States (Cavanaugh et al. 1992).

Similarly, critics of the EU worry that its leveling of the playing field
for trade competition also threatens to level cultural and political differences among nations. Denmark, for example, found that free trade made it
impossible to maintain a sales tax rate higher than neighboring Germany's,
because Danish citizens could simply evade the tax by purchasing goods in
Germany and bringing them across the border duty free. Competitiveness
pressures also make it difficult for a nation to adopt policies that impose
costs on business when low trade barriers force firms to compete with those
in other countries that do not bear such burdens. For example, French firms
demand a level playing field in competing with Spanish firms whenever the
French government mandates employee benefits, health and safety rules, or
environmental regulations more costly than those in Spain. In fact, free
trade tends to harmonize many national policies.

Some trade barriers are designed to protect unique aspects of the economic, social, and political life of nations, especially when trade affects
cultural matters of symbolic importance. For example, France imposes limits on the percentage of television programming that can originate abroad,
allegedly in defense of French language and custom. The obvious target of
these restrictions, U.S. producers of movies and youth-oriented music, contend that the French are simply protecting their own inefficient entertain-
ment industry. Indeed, Hollywood sees *Baywatch* as a valuable export commodity that deserves the same legal protection abroad that the foreign
television sets and CD players that display these images receive in the

United States. But if we restrict trade because we oppose child labor or rain forest destruction, how can we object when other countries ban the sale of U.S. products because they violate *their* values—such as rock music and Hollywood films that celebrate sex, violence, and free expression of controversial ideas or even blue jeans, McDonald's hamburgers, and other symbols of U.S. cultural domination?

■ CONCLUSION:
CHOICES FOR NATIONS AND INDIVIDUALS

Few would deny the contention of liberal theory that trade permits a higher level of aggregate consumption than would be possible if consumers were prevented from purchasing foreign products. It is hard to imagine modern life without the benefits of trade. Of course, it does not follow that trade must be utterly unrestricted because the aggregate economic effect tells only part of the story. As mercantilists remind us, trade also carries with it important social and political implications. Trade shapes the distribution of income and wealth among individuals, affects the power of states and the relations among them, and constrains or enhances the ability of both individuals and nations to achieve goals built on other values. Thus, trade presents a dilemma for nations: No policy can avoid some of trade's negative consequences without also sacrificing some of its benefits. That is why most governments have sought to encompass elements of both liberalism and mercantilism in fashioning their trade policies. The same is true for individuals, because every day each individual must—explicitly or implicitly—assume a stance on the dilemmas identified in this chapter. In turn, trade forces individuals to consider some of the following discussion questions, questions that require normative judgments as well as a keen understanding of the empirical consequences of trade. We must always remember to ask not only what trade policy will best achieve our goals but also what our goals should be.

■ DISCUSSION QUESTIONS

1. Are your views closer to those of a liberal or a mercantilist?
2. Is it patriotic to purchase domestic products? Why or why not?
3. Does one owe a greater obligation to domestic workers and corporations than to foreign ones?
4. Should one purchase a product that is cheap even though it was made with slave labor or by workers deprived of human rights?
5. Should a country surrender some of its sovereignty in order to receive the benefits of joining the WTO?

6. Should one lobby the U.S. government to restrict the sales of U.S. forestry products abroad because these products compromise the environment?

■ SUGGESTED READINGS

Easterly, William (1999) "The Lost Decades: Explaining Developing Countries' Stagnation 1980–1998." Washington, DC: World Bank Policy Research Working Paper.

Fallows, James (1993) "How the World Works," *Atlantic Monthly,* December.

Moon, Bruce E. (1998) "Exports, Outward-Oriented Development, and Economic Growth," *Political Research Quarterly* (March).

——— (2000) *Dilemmas of International Trade.* Second edition. Boulder, CO: Westview Press.

Polanyi, Karl (1944) *The Great Transformation.* New York: Farrar & Reinhart.

Ricardo, David (1981) *Works and Correspondence of David Ricardo: Principles of Political Economy and Taxation.* London: Cambridge University Press.

Smith, Adam (1910) *An Inquiry into the Nature and Causes of the Wealth of Nations.* London: J. M. Dutton.

World Bank (1999) *World Development Indicators 1999.* Washington, DC: World Bank.

7

INTERNATIONAL
CAPITAL FLOWS

Gerald W. Sazama

As humans have developed technology, our race has simultaneously developed economic systems. Primitive economic exchange occurred through the barter of handmade goods within a clan or between neighboring tribes. Modern economic exchange involves the transfer of simple products made with complex machinery, such as wheat, or of the complex machinery itself. These modern exchanges are financed by money and other sophisticated financial instruments. In finance, our race has moved from simple forms of money like wampum beads and gold coins to paper money and checks, and then to international markets for investment funds and national currencies transferred via the Internet.

Like the growth in humans' ability to develop technology, the increasing sophistication of economic systems and financial instruments is wonderful. This sophistication makes possible the complex movement of products and resources first within single countries and now around the globe. Like the development of technology, however, the growth of international capital flows can also be used in horrible ways: for example, for selfish accumulation or for capital flight that dries up a nation's savings base, pushing it into a depression.

Topics discussed in this chapter are (1) a basic vocabulary for international finance; (2) capital flows between industrialized or more developed countries (MDCs), such as the Japanese building a Honda plant in Ohio and the United States having a Chase branch bank in London; (3) capital flows between the MDCs and the less developed countries (LDCs); and (4) an exploration of the question "To regulate or not to regulate the international financial system?"

■ A BASIC VOCABULARY FOR INTERNATIONAL FINANCE

Financial capital takes many forms such as stocks, bonds, loans, and money. During a financial crisis, there is a rapid decrease in the value of financial capital, for example a stock market crash. Financial capital has little or no value if it represents ownership of a broken machine or a dysfunctional corporate organization or if investors lose trust in the health of markets. But financial capital is necessary because it facilitates exchange and movement of real resources like machinery, labor, and technology.

International finance involves financing exports and imports and the movement of financial capital around the globe. To facilitate these flows, we have *foreign exchange markets* in which we exchange the money of one country for that of another. The price at which one country's money exchanges for another's is called the *foreign exchange rate*. An example of a foreign exchange rate would be the number of pesos (the money of Mexico) that can be purchased (exchanged) for one U.S. dollar.

International financial capital flows occur because banks make international loans; multinational corporations (MNCs) and others invest in foreign countries; immigrants send labor earnings to relatives in their home countries; nongovernmental organizations, for example Save the Children Federation, transfer donations; multilateral organizations, for example the World Bank, have loans or grants; and governments build overseas military bases or give foreign aid.

■ NORTH-NORTH CAPITAL FLOWS

While only 15 percent of the world's population live in MDCs, over 80 percent of the world's capital flows among these countries. This is because about 80 percent of the world's production of goods and services is produced in the MDCs (World Bank 2001b). The history and contemporary situation of these North-North capital flows (among the MDCs) are discussed in this section. In the following sections, I will examine capital flows between the MDCs and LDCs, which are much poorer and contain the vast majority of the world's population.

■ *Historical Stages of the Growth of International Finance*

During the nineteenth century, international finance was based on the gold standard, in which all countries agreed to exchange their national currencies for gold. The European nations sent capital to finance the development of emerging MDCs, such as the United States, or to their colonies to build the facilities necessary to extract mineral and agricultural products, which were then sent back to the colonizing country.

Economic competition among the MDCs was an important cause of World Wars I and II. After World War I, Russia became communist. Also, England and France demanded large war reparation payments from Germany, which depressed the German economy and contributed to the rise of Adolf Hitler. In the early 1930s, countries also began to competitively devalue their currency in relation to gold in order to make their exports to other countries cheaper and imports from other countries more expensive. These "beggar-thy-neighbor policies" made an international system of fixed exchange rates with gold unsustainable. The international economy collapsed and with it the use of the gold standard. These factors contributed to the worldwide Great Depression and to World War II.

The United States emerged from World War II as the world's dominant economic power. In the early 1950s, the total gross domestic product (GDP, or the value of a country's output of goods and services) of *all* the MDCs outside the United States was equal to only 40 percent of U.S. GDP. Because of this, the United States had a very strong influence over the terms of international trade and finance during the early postwar period. Other countries used government regulation to fix the value of their currency to the U.S. dollar. This means that these governments guaranteed for a long period that a specific number of units of their currency would equal the value of one dollar. This was done to bring stability to the prices of their goods in international trade.

By the late 1960s, many changes had occurred in the international economy. The European countries and Japan had rebuilt their economies. Most of the LDCs were no longer political colonies. The newly industrializing countries (NICs) such as South Korea, Taiwan, and Brazil were emerging on their own. Also, an economically powerful Union of Soviet Socialist Republics (USSR) was supporting the third world liberation movements for its own political-economic reasons.

All of these factors challenged the economic supremacy of the dollar. Before this time, the United States had a more favorable ratio of exports versus imports. This permitted the United States to fix the redemption of $35 for an ounce of gold. But with the increased international competition, the United States was no longer able to maintain this fixed exchange rate. In 1973, the United States allowed the value of the dollar to fluctuate (or float) in its relation to other countries' currencies. The forces of demand and supply in world markets now determined the relative value of the dollar to foreign currencies and to gold. In response, most other countries also allowed their exchange rates to fluctuate. Adjustment problems to these changes created the first post–World War II international financial crises.

In the 1980s, national governments accelerated the deregulation of banks and international finance. Although this deregulation contributed to a rapid increase in growth of international capital flows, many fear it also led to instability in the system (Blecker 1999; Korten 2001).

The first post-1980 crisis was the third world debt crisis in 1982. This is discussed in the North-South section below. The second post-1980 crisis resulted from the U.S. stock market crash in 1987, which caused a flight of foreign capital from the United States and thereby destabilized other markets. This destabilization was stemmed only with coordinated action by important central banks around the globe. In 1995, there were two additional crises, when the value of the Mexican peso collapsed and when Japanese land values and its stock market crashed. Both again created serious adjustment problems for the international financial system. These problems were minimized only by a series of emergency ad hoc agreements among key national government finance ministers and by the U.S. government's willingness to lend billions of dollars of U.S. taxpayer money to support the peso in order to protect the investments of U.S. corporations in Mexico. The fourth post-1980 crisis, called the Asian crisis, occurred in 1997 and 1998. There was a dramatic sell-off of stock—first in the Thailand stock exchange and then in the Hong Kong stock exchange. Both of these led to drops in stock values in the major international financial markets. The contagion spread, and it resulted in recessions or depressions in several of the Asian countries, including South Korea, Taiwan, Indonesia, the Philippines, and Thailand. This in turn spooked international markets, and the contagion spread as far as causing capital flight from Russia and Brazil. Unfortunately, these local recessions resulted in increased unemployment and reduced government revenues. With lower government revenue, and international pressures to reduce inflation, government expenditures were reduced. Unfortunately, many of these reductions were for social services. Consequently, a disproportionate part of the burden of these recessions fell onto the shoulders of low-income families.

At the time of this writing, the effect on international finance of the terrorist attacks on the World Trade Center and the Pentagon remain to be seen. However, these attacks should not be thought of as isolated from issues of international finance. Citizens from many Middle Eastern countries are frustrated with U.S. support of nondemocratic regimes in the region. These regimes are seen as assuring the United States ready access to the resources of these countries.

■ Contemporary Flows Among the MDCs

Types of flows. Currently there are four broad types of private capital flows among countries: (1) Multinational corporations carry out *foreign direct investment* (FDI). A multinational corporation is a business with at least one subsidiary or joint-ownership company located in a foreign country. Basically, FDI occurs when a multinational owns part of or makes a loan to its foreign affiliate. For example, when Honda builds a plant in Ohio, this is

FDI. (2) *International loans* are money lent by commercial banks and others to private corporations or governments in another country. For example, when the Chase Bank of New York lends money to enable British Airways to expand its service to Russia, this is a foreign loan. (3) *Foreign portfolio investment* occurs when investors buy stock in a foreign corporation on the stock exchange of that country. For example, when a French mutual investment fund purchases Volkswagen stock on the German stock exchange, this is foreign portfolio investment. (4) Finally, there are *international currency* flows. These flows pay for exports and imports of goods and services and support the other types of capital flows.

Size of flows. There was a phenomenal growth in total net global capital flows in the 1990s. They were $794 billion in 1991 and $2,292 billion in 1997; by 2000 they had increased to $4,324 billion. During 1991–1997, about 90 percent of these capital flows were to the MDCs, and in 2000 this percentage increased to 94 (World Bank 2001a: 37). The 1997–1998 financial crises in some LDCs contributed to this post-1997 relative increase of flows to the MDCs.

Although these global capital flows are large, the vast majority of investment remains in the home country. MDCs hold only 3.3 percent of the total value of all of their assets in foreign countries, and only 11 percent of the total value of their loans are to foreigners (Dobson and Hufbauer 2001: 6).

Surprisingly the United States, which is the largest economy in the world, has received since 1983 more capital from foreign countries than it sent abroad. For example, in the year 2000, the United States received $952 billion from foreigners, and it sent abroad $553 billion, for a net receipt of $399 billion (USCEA 2001: 27). In 2000, the United States received about two-thirds of global capital flows among the big-three economies (United States, Japan, and Germany), compared to about one-fifth in 1992. In 2000, net total capital flows into the United States were forty times greater than U.S. foreign aid to LDCs.

The following paragraphs discuss the size of the different types of global financial flows listed earlier.

Global FDI increased from 160 billion in 1991 to 1,118 billion in 2000. During the first two-thirds of the decade, the percentage going to MDCs decreased from 78 to 63. However, because of the financial crises discussed earlier, the percentage of FDI to MDCs increased to 84 for the year 2000 (World Bank 2001a: 37). In essence, the financial crises made companies leery of investing in the LDCs. While U.S. exports totaled $933 billion in 1998, sales by overseas affiliates of U.S.-based multinationals were more than $2.4 trillion.

One concern about FDI is the relative power of multinational corporations. In 1999, only twenty-three of the 162 countries on the globe had a gross domestic product larger than the value of corporate revenue of the largest global corporation, General Motors. The total revenue of General Motors was larger than the combined GDP of the forty least developed countries. Only seven MDCs had a GDP greater than the combined revenue of the five largest global corporations. (UNDP 2001; Kahn 2000). Figure 7.1 compares the corporate revenue of the largest multinational corporations with the GDP of selected countries. The number in parentheses gives the MNC's rank according to corporate revenue. The number in brackets gives the country's rank according to GDP.

Global data on foreign loans is hard to find, but according to the International Monetary Fund, the total value of foreign loans outstanding in 1989 was $917 billion (IMF 1991). Seventy percent of these loans were among the MDCs. The total value of foreign loans increased 460 percent between 1983 and 1989; correcting for inflation, it tripled in just six years.

International currency flows in 1998 are estimated to be $1.5 trillion *per day* (Blecker 1999: 2). These currencies flow back and forth between countries over a year, so the annual net flows are much less than these daily figures times 365. To put the value of currency flows in perspective, the daily value of imports and exports in 1998 was $33 billion (UNDP 2001: 18). The most important cause of these currency flows is speculation to earn financial profits on the miniscule differences in exchange rates in different countries. Ordinarily, this helps keep the various markets in step with each other. However, financial panics can cause a speculative run on a country's currency (in which investors rush to withdraw their assets), as occurred for some countries during the Asian crisis discussed previously.

■ NORTH-SOUTH CAPITAL FLOWS

Having examined North-North capital flows, that is, flows among the MDCs, I now turn to North-South capital flows between the 15 percent of the world's population in the MDCs and the 85 percent of the world's population in the LDCs. The most important areas of controversy for the LDC capital flows are (1) foreign direct investment, (2) the third world debt problem (LDC problems in repaying previous foreign loans), and (3) foreign aid.

■ *Foreign Direct Investment*

There is a long-standing debate among economists as to whether foreign direct investments from MDCs to LDCs are a cause of development, simply one ingredient of it, or an element that retards development. Before the 1980s, many LDCs held that foreign investment retards and distorts

Figure 7.1   Comparing Corporate Revenue and Country GDP, 1999 (in U.S.$ billions)

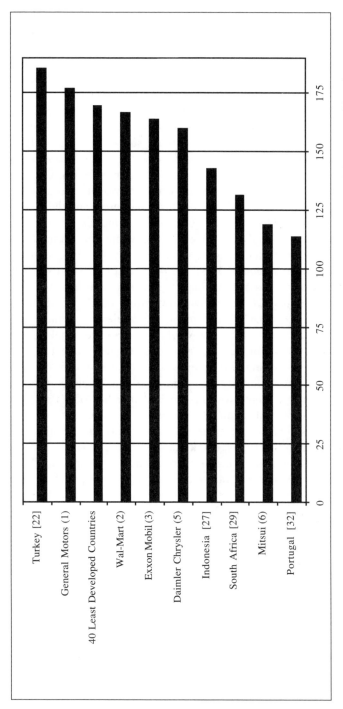

Sources: UN Development Programme, *Human Development Report, 2001* (New York: Oxford University Press, 2001); *Fortune Magazine*, July 24, 2000.

Note: Global rank by country GDP [ ], or by corporate revenue ( ).

economic development. Multinational corporations were accused of extracting raw materials and profits from their host countries, bribing government officials for special privileges, and engaging in cutthroat competition that harms or destroys nationally owned companies. To counter these problems, LDC governments restricted the ability of multinationals to export, import, repatriate capital, set prices, and negotiate with labor unions.

In some instances, distrust of MNCs intensified when MDC governments intervened to protect MNCs from their countries. In 1954, for example, the U.S. government gave military support to Guatemalan rightist forces that overthrew the reformist president, Jacobo Arbenz. As another example, when Salvador Allende was elected president of Chile in 1970, U.S. multinationals feared that his government would expropriate their investments in Chile. In 1972, Allende was murdered, and the U.S. government quickly endorsed the new government, a military dictatorship. This dictatorship was friendly to foreign business, and the United States supported it for almost twenty years.

Other concerns about FDI will be discussed in the last section of this chapter on whether or not to regulate the international financial system.

Proponents of FDI argue that flows from MDCs into LDCs are the missing element in LDC development. They believe that international financial institutions, such as the World Bank and regional development banks, are set up to provide LDCs with capital when and where private investors fail to do so. FDI advocates also argue that most MNCs pay their workers well, negotiate fairly with unions, do not cheat on taxes, and refrain from bribery. There is even some evidence that MNCs pay wages, on average, twice as high as local businesses and offer significantly more fringe benefits: housing, education, hospitalization, and health services.

Beginning in the 1970s, countries such as the "four dragons" of East Asia (South Korea, Taiwan, Hong Kong, and Singapore) sought to make their investment climate attractive. These investment policies helped economic development surge in these countries into the 1990s. Due to a variety of factors, by the mid-1980s, most other LDCs began developing a more favorable climate for FDI as well. One factor in this shift was the failing Soviet Union that had advocated very strict controls on foreign investment. In addition, many LDCs were deeply in debt and corruption-ridden and had state industries unable to face world competition.

Size of the flows. In the early 1990s, private capital flows from MDCs into LDCs suddenly surged. Table 7.1 shows the aggregate net long-term flows of investment capital into LDCs from 1990 to 2000. Official capital denotes capital from governments and international agencies. Note that, while official capital flows remained approximately the same during the decade, private investment increased more than threefold between 1990 and 1997. This increase is spread among all the types of financial flows.

Table 7.1 Net Financial Flows to All Developing Countries, 1990–2000
          (in U.S.$ billions)

	1990	1997	1999	2000
Net capital flows				
Official				
Grants	28.5	24.7	29.2	29.1
Loans (net)	27.2	15.8	16.5	8.5
Total offical	55.7	40.5	45.7	37.6
Private				
Debt (net)	15.6	97.0	−0.6	31.3
Direct investment	24.3	172.6	185.4	178.0
Portfolio equity	3.7	30.2	34.4	47.9
Total private	43.6	299.8	219.2	257.2
Total net capital flows	99.3	340.3	264.9	294.8
Other flows				
Interest on long-term debt	−54.6	−87.2	−100.3	−108.2
Profit remittances on direct investment	−17.6	−31.8	−40.0	−50.2
Total other flows	−72.2	−119.0	−140.3	−158.4
Net financial flows	27.1	221.3	124.6	136.4

Source: World Bank, *Global Development Finance* (Washington DC: World Bank, 2001).

And yet the contagion effects of the 1997–1998 Asian financial crisis caused private capital flows to LDCs to drop by more than 25 percent between 1997 and 1999. In 2000, there was some increase beyond the 1999 level.

Private capital flows have been concentrated in about ten countries, most of which are middle-income countries in East Asia and Latin America and also in two large low-income countries, China and India. The top ten of the 112 developing countries received 74 percent of the FDI to developing countries in the year 2000. In that same year the fifty-eight lowest per capita income developing countries received only 7 percent of FDI to developing countries (World Bank 2001a: 38). Also, there was increased competition for these funds from the former communist countries in Europe. Table 7.2 contains data on the regional disbursement of capital flows to LDCs for the year 2000.

While financial flows from the MDCs to the LDCs help the economic development of the LDCs, these flows also cause flows out of the LDCs back to the MDCs. Multinational corporations remit profits back to their home country, and interest needs to be paid on past debts. As can be seen in the "Other flows" rows in Tables 7.1 and 7.2, these flows back to MDCs are sizeable. For example, Table 7.1 shows that in the year 2000 net private and official flows *to* the LDCs were $294.8 billion, while flows *from* the LDCs were $158.4 billion. Consequently, the net flow to the LDCs was $136.4 billion.

Table 7.2 Net Financial Flows to Developing Countries, by Region, 2000 (in U.S.$ billions)

	East Asia & Pacific	Europe & Central Asia	Latin America & Caribbean	Middle East & North Africa	South Asia	Sub-Saharan Africa	Total
Net capital flows							
Official	11.3	9.6	0.4	1.5	4.0	10.7	37.6
Private							
Debt (net)	-5.0	10.9	15.9	2.4	6.9	0.3	31.3
Direct investment	58.0	28.8	76.2	4.5	3.2	7.3	178.0
Portfolio equity	28.6	5.5	9.9	0.9	2.1	0.8	47.9
Total private	81.6	45.2	102.0	7.8	12.2	8.4	257.2
Total net capital flows	92.9	54.8	102.4	9.3	16.2	19.1	294.8
Other flows							
Interest on long-term debt	-27.7	-17.9	-46.8	-7.2	-5.2	-3.5	-108.2
Profit remittances on direct investment	-16.5	-3.2	-22.4	-2.3	-0.6	-5.2	-50.2
Total other flows	-44.2	-21.1	-69.2	-9.5	-5.8	-8.7	-158.4
Net financial flows	48.7	33.8	33.2	0.2	10.3	10.5	136.4

Source: World Bank, Global Development Finance, Vol. 2 (Washington DC: World Bank, 2001).

■ The Debt Problem

Causes. When the Organization of Petroleum Exporting Countries (OPEC) raised the price of oil in 1973, the impact was felt primarily in the then non-oil-producing less developed countries (sometimes known as NOPEC). While MDCs had much larger increases in their oil payments, they did not feel the same effects, since OPEC mostly invested its windfall proceeds back into the MDCs.

To sustain its development programs NOPEC received special loans from the MDCs. Since nothing had happened to increase NOPEC exports or to attract increased investment, their indebtedness piled up, marking the beginnings of the debt crisis.

Strangely, the debt crisis spread to oil-producing countries as well. On the strength of further revenue prospects, these countries increased their borrowing for "development projects." But no great increase in development was realized in many of these countries, possibly due to wasteful projects and corruption. Then when oil prices tapered off in the 1980s, they were unable to pay their debts.

Third world debt has continued to increase over time. One cause contributing to this increase has been the high interest rates in the MDCs, particularly in the 1980s. For a loan with a compound interest rate of 10 percent, for example, the debt doubles in about seven years, even if no new loans are taken out. The indebted countries, for their part, have been eager to borrow more to sustain their domestic consumption and investment, to alleviate balance-of-payments deficits, and to repay old loans.

Extent of the debt problem. Between 1980 and 2000, outstanding LDC foreign debt increased from $587 billion to $2,528 billion, or from 18 to 37 percent of LDC GDP (World Bank 2001c: 246). On average for all LDCs in 1999, debt repayments were 5.8 percent of GDP and 22.3 percent of revenue from exports. As can be seen in columns 7 and 8 of Table 7.4, these debt burden indices varied substantially from region to region.

The World Bank has developed lists of countries that have high debt burdens. The LDCs are first divided into low income (1998 per capita income less than $761), lower-middle income (1998 per capita income between $761 and $3,030), and upper-middle income (1998 per capita income between $3,031 and $9,360). Of the sixty-three low-income LDCs, thirty-three are classified as highly indebted low-income countries, or HILICs. Ten of the fifty-seven lower-middle-income LDCs and three of the thirty-six upper-middle-income countries have been classified as highly indebted (World Bank 2001c). Debt is relatively most severe in African countries south of the Sahara, where the standard of living and degree of economic development are among the lowest in the world.

Debt restructuring. For reasons that combine political interests, compassion, and the need for world stability, industrialized nations, the World Bank, the International Monetary Fund (IMF), and other international agencies have been seeking ways to reschedule and reduce third world debt. In the early 1980s, some key banks, such as the Bank of America, held a substantial percentage of their loan portfolio in loans to LDCs. When many countries found that they could not repay their loans, these large banks in the MDCs wrote off part of this debt as bad loans. These write-offs created a global financial crisis that contributed to the recessions of the early 1980s in both the United States and Europe.

Given these problems, the international financial community has sought ways to restructure the economies of indebted countries in order to contain increases in indebtedness. As a condition for its assistance to indebted LDCs, the IMF demanded that these countries reduce market regulations and have exchange rates that according to the IMF reflect the free market demand and supply for these countries' currencies in the global financial markets. As time went on, privatizing inefficient state enterprises and balancing government budgets were added to the list. Such conditions for refinancing are called conditionality.

Conditionality issues. IMF conditionality, also required by the World Bank and governments of MDCs, is very controversial. Many view these conditions as necessary for the increased economic growth needed by LDCs to repay their debts. Others, however, hold that conditionality is big banks from rich MDCs interfering in the internal politics of LDCs.

In the LDCs, the public complains that the poor suffer from the resulting austerity, since unemployment increases and government services are reduced. Public protests, demonstrations, and riots have resulted. The IMF argues that the suffering is inevitable due to wasteful government policies and that the IMF program is actually diminishing suffering by promoting less wasteful policies. In addition, some economists point out that elites in the developing countries have squandered the borrowed funds and the poor are now suffering from the austerity needed to bring order into government accounts. Recent changes to conditionality are discussed in Chapter 8.

The Jubilee 2000 movement is an effort begun in 2000 by religious organizations and other civic groups to urge debt forgiveness, especially for the HILICs. Partially as result of these efforts, between 1996 and 2000, the World Bank and the IMF committed total debt relief of $33.6 billion. Of this, $30 billion was committed in 2000 (World Bank 2001c: 100). Private banks have been more reluctant to negotiate such relief. This $33.6 billion of debt relief is 1.3 percent of the total outstanding LDC debt in the year 2000. However, relief was focused on the debt of the highly indebted low-income countries.

It is not necessarily easy for a private financial institution to write off (forgive) debt. This means it will have lower profits or even experience losses. Corporate executives have their jobs on the line according to the profitability of their company. When you invest your pension funds, would you be willing to have slower or even negative growth in the value of your stocks? Also, these institutions argue that internal problems in the HILICs contribute to their inability to repay. Why should lenders shoulder the whole responsibility?

Those in the Jubilee 2000 movement argue that there is no way that the HILICs can repay all of these debts. Efforts to do so will only result in tremendous burdens on the poor in these countries. This is seen as a religious question of social justice. Also, the MDCs' institutions should share the burden of past mistakes. They agreed to make the loans. Finally, orderly write-offs of bad debts will contain the contagion effects of future financial crises.

The Jubilee 2000 movement continues. It brings attention to the plight of the poor in the LDCs, and it has resulted in the write-off of some debt. More information on this question can be found in the readings and at the websites listed at the end of this chapter.

Foreign Aid

Two-thirds of foreign aid is bilateral (country to country); the remainder is multilateral (countries giving aid via international agencies such as the World Bank and UNESCO, the United Nations Educational, Scientific and Cultural Organization) (World Bank 2001b). About 70 percent of foreign aid is grants, aid that does not require repayment (World Bank 2001b). The balance comes as concessionary loans. Concessionary loans are subsidized via lower interest rates and longer repayment periods than private sector loans.

Nongovernmental organizations (NGOs), such as Save the Children Federation and the International Red Cross, also give aid to LDCs. They give aid that they collect from private donations, and increasingly governments are choosing to give some of their foreign aid through the NGOs. In 1998, 28 percent of foreign aid passed through NGOs, 40 percent of which came from government funds and the balance from private donations. This increase in governmental funds channeled through NGOs provides NGOs with new opportunities, but many are worried that governments will increasingly set the agendas of these groups.

Also, foreign aid comes in disguised form, such as special permits for an LDC to export to an MDC. For example, European countries try to give special preferences for bananas exported by former colonies in Africa.

To put this into perspective, while total Official Development Assistance (ODA) in 2000 was $37.5 billion, remittances back to the home country LDCs from immigrants working in MDCs was $56 billion (World Bank

2001c). ODA includes the grants and concessionary loans discussed above plus nonconcessionary loans (given on market terms).

Size of foreign aid flows. Table 7.3 shows ODA by the principal donor countries. The Development Assistance Committee (DAC) is an organization of the twenty-two largest MDCs. It gives about 90 percent of all foreign aid. Most of the rest comes from Arab countries (World Bank 2001a). The United States gives the second largest total amount of ODA, but it gives the *lowest* percentage of its gross domestic product. Among the DAC countries, the United States had the largest decrease in ODA as a percentage of GDP between 1965 and 1999.

Total ODA from DAC countries in 1999 was $56.4 billion, an average of 0.24 percent of DAC countries' GDPs. In the 1970s, some international agencies and NGOs urged MDCs to increase foreign aid to 1.0 percent of the GDP. However, from Table 7.3 we see that on average it decreased by 0.23 percent of GDP between 1965 and 1999. In 1994, the United Nations Development Programme (UNDP) estimated that if the global economy directed $40 billion more ODA per year for the next ten years to basic human needs, this would provide universal access to safe drinking water, primary health care, basic education, and a basic family planning package to all willing couples (UNDP 1994).

Table 7.4 shows ODA by recipient regions. The sub-Saharan African countries receive the most per capita *and* total dollars of foreign aid. Also, ODA is the highest percentage of their GDPs.

Table 7.3   Official Development Assistance, by Principal Donor Countries, 1999

	% of GNP 1999	U.S.$ Billions 1999	% of GNP 1999 minus % of GNP 1965
Denmark	1.01	1.7	n.a.
Netherlands	0.79	3.1	n.a.
France	0.39	5.6	−0.35
Japan	0.35	15.3	0.09
Canada	0.28	1.7	0.10
Germany	0.26	5.5	−0.15
United Kingdom	0.23	3.4	−0.19
United States	0.10	9.1	−0.49
Twenty-two DAC countries[a] average (total)	0.24	(56.4)	−0.23

Sources: UNDP, *Human Development Report, 1996* and *2001* (New York: Oxford University Press, 1996 and 2001).

Notes: n.a: not available.

a. DAC is the Development Assistance Committee of the Organization for Economic Cooperation and Development.

Table 7.4 Comparative Measures of Official Development Assistance, Private Flows, and Debt Repayments, by Region, 1999

	(1)	(2)	(3)	(4)	(5)	(6)	(7)	(8)
	Per Capita GDP (U.S.$)[a]	Official Development Assistance			Net FDI[b] as % of GDP	Other Private Flows as % of GDP	Debt Repayments	
		Per Capita (U.S.$)	U.S.$ Billions	As % of GDP			As % of GDP	As % of Exports
Forty least developed countries	1,170	17.8	10.6	7.0	3.0	−0.1	2.8	13.0
All developing countries	3,530	7.2	33.0	0.6	2.9	0.4	5.8	22.3
Eastern Europe & Central Asia	6,290	18.6	7.4	0.8	2.9	0.9	5.1	16.5
Developing countries by region								
Arab states	4,550	18.3	4.3	0.8	0.3	0.3	3.6	11.4
East Asia & Pacific	3,950	4.9	8.9	0.5	3.0	−0.2	15.7	15.8
Latin America & Caribbean	6,880	9.2	4.5	0.2	4.5	1.1	23.6	41.6
South Asia	2,280	3.1	4.3	0.6	0.5	−0.3	2.8	16.6
Sub-Saharan Africa	1,640	18.3	11.0	3.6	2.4	0.8	4.6	14.3

Source: UNDP, *Human Development Report, 2001* (New York: Oxford University Press, 2001).
Notes: a. The per capita GDP of the United States is $31,872. It is $26,050 for the twenty-three highest per capita income countries.
b. FDI (foreign direct investment).

Note that the regional distribution of ODA is quite different from the distribution of private flows (compare columns 4, 5, and 6 in Table 7.4). This is because private capital goes to places where the best profits can be earned, and foreign aid is given for motives that are more complex.

Arguments for giving and receiving foreign aid.   Reasons *for giving aid* include the following: (1) political, for example an MDC supports a particular LDC government that has policies favorable to the granting MDC; (2) military, for example to economically support military allies in a region of conflict, such as the United States giving aid to Central America in the early 1980s and aid to Afghanistan as a result of the war on terrorism (note: the ODA figures in the preceding section and tables do not include direct military aid); (3) economic, for example to open markets in the LDCs for the donor country's businesses; (4) sociopolitical, for example U.S. aid to Colombia, as part of the U.S. drug enforcement and military policies; (5) humanitarian, for example to prevent malnutrition and diseases and to provide basic needs for all. A larger humanitarian reason would be to redistribute resources from the prosperous MDCs to the poor LDCs in order to alleviate the maldistribution of global output, thereby encouraging peace and democracy throughout the globe.

The following comments will help us to understand the relative strength of these various motives for giving aid. One explanation for the relative decrease in foreign aid between 1965 and 1999, especially for the United States, is the end of the Cold War with the former USSR. During the Cold War, aid was given to bring countries into one's so-called sphere of influence. Global figures on military aid are not readily available. However, for the United States in 1996 military aid was 40 percent as large as development aid (Hoy 1998: 22). As an example of aid for economic motives, the United States sent $280 and $176 per person in poverty, respectively, to Egypt and Israel; both are key players in the oil-rich Middle East. Meanwhile the United States gave $19 per poor person to Bangladesh, one of the poorest countries on the globe (UNDP 1994).

As for humanitarian aid, 7 percent of total bilateral ODA goes directly for basic needs, as does 16 percent of multilateral ODA (UNDP 1994). Some humanitarian aid has now changed its form into debt relief for the highly indebted low-income countries. This will be discussed in the following section. In the summer of 2001, the Group of 7 (G7) countries pledged $1.3 billion for AIDS relief in Africa, but the United Nations says $7–10 billion will be needed per year. Ironically, at the same time the United States is pressuring LDCs to remove import restrictions on cigarettes, even though they kill more people than AIDS (Fairclough 2001: 1).

The principal argument *for receiving aid* is economic. Foreign aid helps a country acquire the capital needed for economic development. If

outside funds are invested wisely, they create jobs and output. This new income can then be used to repay the loans. Indeed, according to this view,
once a country becomes sufficiently developed it becomes a capital exporter.

Arguments against giving and receiving foreign aid. Arguments *against giving
aid* include the following: (1) It is inappropriate to interfere in another
country's internal affairs. (2) Countries need to get their own houses in
order before seeking aid, otherwise the aid is wasted or, worse, it fosters internal corruption, goes to the elite, and worsens the conditions of the poor.
(3) Money would be better spent on the poor at home—why send it abroad?
However, it should be noted, if all foreign aid were spent on domestic social welfare expenditures in the MDCs, these expenditures would increase
from 15 percent of their GDPs to 15.3 percent (UNDP 1994).

Some arguments *against receiving aid* are: (1) Countries should have
the right to choose their own style of economic development, and foreign
aid skews these choices to the visions of the grantor. The vast majority of
all foreign aid, for example, goes for large-scale modernization projects.
Also, some do not accept what they call an implicit assumption in foreign
aid, that all countries should become industrialized and that they should become politically like the current MDCs. (2) Foreign aid creates enclaves
(isolated pockets) of international mining, agribusiness, and unskilled manufacturing. Because of their geographic or economic isolation, these enclaves have little positive influence in helping modernize the domestic
economy. Rather, they exist to export national resources to the MDCs. (3)
Much foreign aid is "tied" aid, meaning that the aid has to be spent in the
granting MDC. (4) A substantial amount of aid is given as loans, creating
future indebtedness for a country. (5) Receiving aid can be a bureaucratic
nightmare, because of the problems of inconsistency among donor policies
and administration of many small projects. For example, in the early 1990s
in Tanzania, there were forty donors and more than 2,000 projects (World
Bank 2001c). (6) Some argue that even humanitarian aid given at times of
famine results in only a temporary fix. In the long run, such aid harms the
internal adjustment process necessary for a long-term solution.

Meanwhile, some argue "trade, not aid." On the one hand, some MDC
countries like the United States say that LDCs need to remove government
restrictions on business because they retard the growth of export industries.
If the LDCs did this, they would not need so much foreign aid. On the other
hand, many LDC governments claim that protectionism in the MDCs de-
prives them of export markets important for their development. With 85 percent of the world's population, the LDCs had only about 27 percent of world
exports in 1998. Again, this trade was concentrated in the more prosperous
LDCs. A related question is the long-term decline in the relative prices of LDC
exports. Between 1980 and 2000 prices on non-oil commodities exported by

LDCs dropped by 50 percent, and oil export prices dropped 43 percent (World Bank 2001b: 229). So LDCs had to sell more for less just to try to stay in place.

■ TO REGULATE OR NOT TO REGULATE
THE INTERNATIONAL FINANCIAL SYSTEM?

The principal contemporary question for international finance is to regulate or not to regulate? Essentially the way people and governments answer this question depends on their trust in private enterprise, or the capitalist system. Classical economic liberals believe that ordinarily free markets result in the maximum benefit for all. Therefore, they want little or no regulation. Traditional political liberals, in contrast, believe that free markets usually result in maximum benefit for all but that at times free markets cannot take care of all of society's needs; free markets, according to this perspective, can even be harmful to society. Therefore, traditional liberals would be willing to have more regulation of a market economy. "Activists" tend to have much less faith in a free market economy. Rather, they believe that free markets left on their own can give benefits but also can create deep problems for society. Therefore, activists favor substantial regulation, and they urge that this regulation should be democratically controlled.

Before proceeding, it should be noted that putting labels on schools of political economic thought is a tricky matter. Frequently, groups taking stances on a specific issue are an amalgam of a variety of positions. Therefore, on an international level "neoliberals," who are liberal in domestic politics, have often supported the laissez-faire (free market) policies of the classical liberals. Meanwhile, traditional political liberals (that is, liberals in the sense of U.S. politics) have usually supported some government regulation of private markets. But as the economy has changed and politics have shifted, many in this coalition now "pick and choose" where and where not to regulate. Finally, the activists are a coalition of a broad spectrum of groups opposed to the global economy as it is currently developing. Some are interested in protecting jobs, others the environment; others seek social justice; still others favor a return to simpler, decentralized, semiautonomous economies. Some are nonviolent, and others believe that violence toward property is an appropriate strategy. These activists are parts of the diverse protests that have occurred in Seattle, Genoa, Italy, and Washington, D.C., and other cities.

Understanding these general policy orientations will help you develop your answer to the question "To regulate or not to regulate the international financial system?" But first I will address just what governments might want to regulate. Then I will consider just who might do the regulation.

Finally, with this framework, I will examine the arguments for and against regulation.

■ What Could Be Regulated?

The North-North section earlier in this chapter explained the four types of international capital flows. Each of these flows occurs for different purposes, and some types can be withdrawn from a country more quickly than others. Therefore, discussion of whether to regulate and how to regulate can differ for each type.

■ Who Should Do the Regulation?

Those favoring minimal regulation tend to favor placing the responsibility for regulation on national governments. Governments should keep their economy in order by eliminating corruption and assuring efficient markets. Also, they do not want their country to surrender any national autonomy to an international agency.

Those favoring more regulation favor a series of ad hoc international agencies and treaties to take care of specific problems and subsystems of the international economy. In some ways, this is similar to the present system of international regulation. For example, the International Monetary Fund provides emergency loans to developing countries and regulates their foreign exchange markets. The World Trade Organization (WTO) has some regulatory power over exports and imports among member countries to assure adherence to the principle of free trade. There are regular meetings of the presidents, prime ministers, and financial ministers of the largest global economies, called the G7. These leaders coordinate national policies affecting the global economy.

Some oppose this ad hoc regulation because they believe it is too restrictive, so it slows growth and distorts the international economy. Others oppose this ad hoc regulation because they believe it is insufficient to take care of the interrelatedness of global economic problems and not strong enough to solve global financial crises. Still others oppose this system because they believe it is too much under the control of the large corporations and powerful countries.

Those favoring the most regulation favor stronger international agencies. Some would do this by giving more power to existing agencies, and others would create new forms of world government. Under the most far-reaching form of this approach, there would be a world central bank, a common global currency, and a world federation that would regulate the global economy much like the European Union now has a common central bank, a common currency (the euro), and a European parliament.

■ Arguments For and Against Regulation

Arguments for regulation. The principal argument for increased regulation of international financial markets is to avoid the frequency and the costs of the international financial crises that have occurred since 1980. Global financial markets have grown rapidly and are integrated, but the oversight power of existing national government and international agencies has not kept pace with these changes. This results in financial panics on an international level similar to the old-fashioned runs on banks that occurred in the free-market capitalism of the United States from 1880 to 1929. As result of the collapses that occurred during this period, the Federal Reserve Bank was created in 1913. During the Great Depression of the 1930s, it was given additional power, and other regulatory agencies were created.

Some argue that people in the MDCs, especially in the United States, have been shielded from the most severe effects of the recessions and even depressions in the LDCs affected by recent international financial crises. Therefore, people in MDCs have been reluctant to support much increase in international regulation. But this may be shortsighted, because the Great Depression of the 1930s was linked to an international financial crisis that spread from country to country.

A second important argument for international regulation is that unregulated international capitalism results in a "race to the bottom." Unregulated competition forces companies to sell the best product possible for the cheapest price possible. This results in companies moving their plants to countries that allow the lowest wages, oppressive working conditions, and poorly enforced or nonexistent environmental protection laws. The rebuttal to these arguments is given in the following section on arguments against regulation.

I have to admit that recently I found myself in a dilemma related to this issue. Shopping for a dining room table, I found the showroom had one of excellent quality manufactured in Vermont costing $750 and a second of good quality manufactured in Thailand costing $350: a difficult decision. Those favoring regulation say a way out of this dilemma is to have enforceable international labor and environmental standards.

The growth of the antisweatshop movement, and of groups concerned with the global environment, is a response to the worry about a race to the bottom. These groups are concerned with conditions in the LDCs. But they also worry about low standards there dragging down living conditions for workers in the MDCs and environmental quality for the whole globe. The demonstrations in Seattle, Prague, Washington, D.C., Quebec City, and Genoa at meetings of the World Trade Organization, World Bank and International Monetary Fund, the Inter American states meeting on free trade, and G7, together with the growth of related campus-based organizations, show the depth of this concern.

A third argument for regulation is to prevent personal abuse and reckless selfish speculation. The example of the 1995 scandal surrounding a twenty-eight-year-old investor in Singapore is a case in point. This young man sitting at a computer terminal made millions of dollars in profits for his bank, and he earned salary and bonuses of several million dollars for himself. Then, over roughly a four-week period, he speculated with $29 billion of the bank's money and ran up losses of $1.3 billion. Consequently, the venerable 223-year-old Barings Bank of England went bankrupt, and the Tokyo stock exchange index dropped (Korten 1996).

A fourth concern about an unregulated global economy is its assumption that economic growth is good. Does this result in consumption of more than we need, coupled with anxious working conditions with long hours? Would regulation that distributes from the "haves" to the "have-nots" reduce the global race for more?

Arguments against regulation. The principal argument against additional regulation of the international financial system is essentially, "If it's not broke, don't fix it." The existing institutions have worked well. The incredible growth of international private capital flows since 1980 is proof that the current system is working. Also, the damage from regional crises generally was stopped from spreading. Some countries were affected by financial contagion, but this was due to their own internal weakness, such as corruption and cronyism, bad loans, government deficits, and artificial controls on business. Also, because of their own lack of internal democracy, national elites put much of the costs on the poor.

A second argument against regulation is that free markets have resulted in a substantial increase in the general standard of living in many LDCs. This is part of the rebuttal against the race-to-the-bottom argument given in the previous section. Per capita income in the higher-income LDCs, which contain 49 percent of the world's population, increased by 2.8 percent per year for the 1980–1996 period. In comparison, for the MDCs, which contain 16 percent of the world's population, per capita income only increased by 1.4 percent per year for the same period (World Bank 1999). Given the right conditions, this economic development can spread to the 39 percent of the world's population living in the poorest LDCs, whose per capita income increased at only 1 percent per year in the 1980–1996 period.

This economic development will create pressures for more democracy, better workers' organizations, and a demand for better environmental quality. Indeed, according to the United Nations Development Programme, the percentage of the world's population living in countries with a low Human Development Index has dropped from 33 percent in 1970 to 10 percent in 1999 (UNDP 2001: 11). Also, a World Bank report shows a decrease in urban air pollution in São Paulo; Mexico City; and, on average, for urban areas in China, while there was a simultaneous increase in FDI in these locations (World Bank 2001b: 72).

A third argument against regulation is that if investors or corporations are reckless or inefficient, the market will discipline them. Thus the man from Singapore lost his job, and his bank went under. The price they paid is a lesson for others. If we put in place international regulation, it will just end up paying the costs of those who made mistakes with taxpayers' money. Consequently, those against regulation argue that others will behave recklessly, knowing that they will not personally suffer the consequences of their actions.

A fourth argument against regulation is that when a multinational invests in a foreign country, it brings new technology and new management and marketing styles, provides jobs, and increases that country's output.

A fifth argument against regulation is the belief that the free market system has made life easier for many of us in the MDCs. Few of us wish to return to the backbreaking work and low life expectancies of the nineteenth century. But most citizens of the LDCs still experience this backbreaking work and low life expectancy. They want economic development. Those against regulation say this requires both free markets and time for the changes to occur.

We are left with a series of questions that are not easily answered. Should we regulate? If so, how? Can a balance be struck between international regulation and national autonomy? If we use regulation, will large countries, large corporations, and special interests control it for their own benefit? Many, including myself, believe markets need to be open but responsible. To ensure responsibility some form of democratically controlled regulation is desirable. But that is not easy to achieve.

■ CONCLUSION

Along with modern technology, we humans have invented our international financial institutions, and like our use of technology, we can use these economic institutions for good or for ill. International capital flows facilitate the worldwide movement of goods, services, and savings. These flows can facilitate global economic growth, integration, and justice. If, however, we do not stay in contact with some higher moral principles (which some would call our Inner Light, or God) these flows can be used to gain economic power over others and to create instability and injustice in the international economy. Some argue that global governance is necessary to ensure a stable and just flow of international capital throughout the globe, while others argue that these goals will be better served by financial markets free of regulation.

■ DISCUSSION QUESTIONS

1. Discuss how the global financial system is an institution created by humans, which we can use for good or for ill.
2. What are the implications of the United States being a large net capital importer?
3. On balance, does private foreign direct investment promote or discourage economic development in the LDCs? Discuss strategies that LDCs might adopt to make FDI fit their development aspirations without severely harming incentives for FDI.
4. How would you solve the LDC debt problem, keeping in mind stability of the international financial system as one of your goals? What is conditionality? Is it necessary, or is it oppressive to the LDCs?
6. Should the United States give more or less foreign aid? If more aid, what are appropriate motives for giving aid, and where and for what should the aid go?
7. As foreign aid is becoming less important as a source of capital to the LDCs, what other policies would you suggest to encourage economic development in the more, and in the less, prosperous LDCs?
8. Have classically liberal trade policies, which allow MNCs to move manufacturing plants from country to country, caused a race to the bottom? If yes, how might it be eliminated without slowing growth in the LDCs?
9. What are the advantages and disadvantages of increasing global governance over the international economy?

■ NOTE

I am indebted to John Powelson of the University of Colorado for his assistance with some portions of this chapter.

■ SUGGESTED READINGS

Blecker, Robert A. (1999) *Taming Global Finance*. Washington, DC: Economic Policy Institute.

Dobson, Wendy, and Gary Clyde Hufbauer (2001) *World Capital Markets*. Washington, DC: Institute for International Economics.

Graham, Edward M. (1996) *Global Corporations and National Governments.* Washington, DC: Institute for International Economics.

Hoy, Paula (1998) *Players and Issues in International Aid.* Bloomfield, CT: Kumarian Press.

Korten, David C. (2001) *When Corporations Rule the World,* Second edition. Bloomfield, CT: Kumarian Press.

Nijman, Jan, and Richard Grant (1998) *The Global Crisis in Foreign Aid.* Syracuse, NY: Syracuse University Press.

Reinicke, Wolfgang H. (1998) *Global Public Policy: Governing Without Government?* Washington, DC: Brookings Institution.

Todaro, Michael P. (1999) *Economic Development.* New York: Addison-Wesley. (Chapters 14 and 15 are an excellent introduction to this topic as it affects LDCs.)

UNDP (United Nations Development Programme) (2001) *Human Development Report.* New York: Oxford University Press.

World Bank (2000) *World Development Report 2000/2001: Attacking Poverty.* New York: Oxford University Press. See especially chapter 11 on foreign aid.

——— (2001) *Global Development Finance.* Washington, DC: World Bank.

These Internet sources provide useful information on the topic of international finance:

Bread for the World, www.bread.org

Debt Channel, www.debtchannel.org

Jubilee 2000 follow-up, www.jubilee2000uk.org and www.2000usa.org

One World, www.oneworld.net

Oxfam International, www.oxfaminternational.org

United Nations Development Programme, www.undp.org

World Bank, www.worldbank.org

8

POVERTY IN A GLOBAL ECONOMY

Don Reeves

- Poverty is a mother's milk drying up for lack of food or kids too hungry to pay attention in school.
- Poverty is to live crowded under a piece of plastic in Calcutta, huddled in a cardboard house during a rainstorm in São Paulo, or homeless in Washington, D.C.
- Poverty is watching your child die for lack of a vaccination that would cost a few pennies or never having seen a doctor.
- Poverty is a job application you can't read, a poor teacher in a run-down school, or no school at all.
- Poverty is hawking cigarettes one at a time on jeepneys in Manila or being locked for long hours inside a garment factory near Dhaka or in Los Angeles or working long hours as needed in someone else's field.
- Poverty is to feel powerless—without dignity or hope.

■ DIMENSIONS OF POVERTY

Poverty has many dimensions. Religious ascetics may choose to be poor as part of their spiritual discipline. Persons with great wealth may ignore the needs of those around them or may miss the richness and beauty of nature or great art and remain poor in spirit. But this chapter is about poverty as the involuntary lack of sufficient resources to provide or exchange for basic necessities—food, shelter, health care, clothing, education, opportunities to work and to develop the human spirit.

Globally, poor people disproportionately live in Africa. The largest number live in Asia. A significant number are in Latin American and Caribbean countries. Up to 80 percent of the people in several sub-Saharan Africa countries and Haiti are poor, and more than one-quarter of all people in developing countries together are poor. (The situation worldwide is shown in Figure 8.1.)

But no place on the globe is immune to poverty. The United States, some European countries, and Australia also have large blocs of poor people. With few exceptions, the incidence of poverty is higher in rural than in urban areas but is shifting toward the latter. Nearly everywhere, women and girls suffer from poverty more than do men and boys; infants, young children, and elderly people are particularly vulnerable. Cultural and discriminatory causes of hunger are immense; the difficulties in changing long habits and practices should not be underestimated (BFW Institute 1994).

In this chapter, I look first at ways in which poverty is measured. Then I will look at approaches to reducing poverty in the context of a global economy, especially the relationship between economic growth and inequality. Finally, I will examine a series of policy choices developing country societies might consider as they attempt to reduce poverty.

■ MEASURING POVERTY AND INEQUALITY

Poverty is not the same in the United States or Poland or Zimbabwe. It will often be described differently by supporters or critics of a particular regime. Poverty does not lend itself to an exact or universal definition. Deciding

Figure 8.1 Number and Percentage of Poor People Worldwide, 1990 and 1999

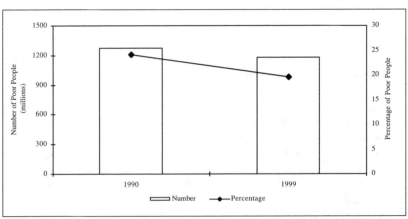

Source: World Bank, *World Development Indicators* (New York: Oxford Univeristy Press, 2001).

who is poor depends on who is measuring, where, and why. So, first, I will describe how poverty and inequality are measured.

■ Poverty Thresholds

Poverty is usually measured by income or consumption. By the most widely used measure, the World Bank estimates that worldwide 1.2 billion people live on incomes equivalent to less than U.S.$1 per day (World Bank 2001b). The majority of these people chronically lack some or all basic necessities. The rest live so close to the edge that any emergency—illness, work layoff, drought—pushes them from just getting by into desperation.

In the United States, poverty is defined as three times the value of a thrifty food plan devised by the U.S. Department of Agriculture, adjusted for family size—$17,650 for a family of four in 2001. Some critics say it is higher than necessary, partly because certain government program benefits— such as Medicare, housing subsidies, and school meals—are not counted. But poor people themselves feel hard pressed. The thrifty food plan was worked out in the early 1960s to address short-term emergencies. Although it is adjusted annually for changes in food prices, other costs, particularly housing, have grown faster than food costs since the plan's base year (1955); so the threshold has represented a gradually declining standard of living.

The selection of the poverty threshold often makes a dramatic difference in the observed poverty rate. The World Bank, based on its cutoff of $1 per person per day, estimates that in 1998, 225 million (18.5 percent) of China's total 1.2 billion people were poor. The Chinese government, using a lower cutoff point, claims that only about 50 million (4.6 percent) of its people were poor in that year (World Bank 2001b).

■ Gross Domestic Product and Gross National Product

Two other widely used income measures are per capita gross domestic product (GDP) and gross national product (GNP). GDP is the value of all goods and services produced within an economy; GNP equals GDP plus or minus transfers in and out of the economy, such as profit paid to foreign investors or money sent home by citizens working abroad.

Among the world's economies, large and small, the World Bank counts sixty-four low-income economies with an annual per capita GNP from $100 to $755 (1999). At the other end are fifty high-income economies with per capita GNP from $9,266 to $38,300. In between are ninety-three middle-income economies with per capita incomes between $756 and $9,265 (World Bank 2001a).

GNP (or GDP) provides a quick measure of the capacity of an economy overall to meet people's needs. It also represents the pool from which savings and public expenditures can be drawn. But GNP is seriously flawed as a measure of poverty or well-being because it gives no information about the quality of the production or the distribution of income within the country.

First, GNP and GDP fail to distinguish among types of economic activity. Manufacturing cigarettes, making bombs, and running prisons are scored as contributing to GNP or GDP the same as making autos, teaching school, building homes, or conducting scientific research. Second, many goods and services generate costs that are not reflected in their prices—polluted air from manufacturing or illness from overconsumption, for example. Third, many nurturing and creative activities—parenting, homemaking, gardening, and home food preparation—are not included because they are not bought and sold. At best, GNP and GDP figures include only *estimates* for food or other goods consumed by producers, unpaid family labor, and a wide range of other economic activities lumped together as the informal sector. Illegal or criminal activities, such as drug-dealing or prostitution, are generally not included in estimates but nonetheless contribute to some people's livelihood.

■ *Purchasing Power Parity*

GNP and GDP figures for various countries are usually compared on a currency exchange basis. The per capita GNP in Bangladesh, at 18,000 taka, could be exchanged for U.S.$370.

But 18,000 taka will buy more in Bangladesh than $370 will buy in the United States, primarily because wages there are much lower. Thus, the World Bank and the United Nations Development Programme (UNDP) have adopted a measure—purchasing power parity (PPP)—that estimates the number of dollars required to purchase comparable goods in different countries. Bangladeshi PPP is estimated at $1,530, rather than $370 (World Bank 2001a).

PPP estimates make country-to-country comparisons more accurate and realistic and somewhat narrow the apparent gap between wealthy and poor countries. Even so, vast disparities remain. PPPs of $24,000 to $32,000 per capita—as in the United States, Switzerland, and Canada—are forty to sixty times those of Tanzania and Ethiopia, at $500 and $620 (World Bank 2001a).

■ *Inequality*

Estimates of poverty and well-being based on estimated GDP are at best crude measures. GNP, GDP, and PPP are all measured as country averages. But because poverty is experienced at the household and individual level, the distribution of national incomes is a crucial consideration.

Detailed and accurate information is necessary for targeting anti-poverty efforts and particularly for assessing the consequences of policy decisions in a timely fashion. But census data as comprehensive as that for the United States are a distant dream for most poor countries. Many of them do not keep such basic records as birth registrations and may have only a guess as to the number of their citizens, let alone details about their conditions. Representative household surveys are the only viable tool for most countries for the foreseeable future.

Household surveys, to be useful—especially for comparison purposes—need to be carefully designed, accurately interpreted, and usable for measuring comparable factors in different times, places, and circumstances. Private agencies, many governments, and even some international agencies are tempted to shape or interpret surveys to put themselves in the best light. Users of survey results need to be keenly aware of who conducted the survey and for what reasons.

Global inequality. Globally, we have accepted gross income inequality. The most used measure of inequality compares the income of the richest one-fifth, or quintile, of each population with that of the lowest quintile. Measured in currency exchange value, the wealthiest one-fifth of the world's people control more than 85 percent of global income. The remaining 80 percent of people share less than 15 percent of the world's income. The poorest one-fifth, about 1.2 billion people, receive barely 1 percent. The ratio between the average incomes of the top fifth and the bottom fifth of humanity is 70 to 1 (see Figure 8.2 and Table 8.1).

Using purchasing power parity as a measure, the poverty gap narrows, but it is still extreme. The first-ever global survey based on household surveys, and using PPP, estimated that during the 1990s the poorest 20 percent could purchase on average about one-twelfth as much as the top 20 percent, and the ratio increased during this period. This contrasts with estimates based on national averages, as reported by the United Nations Development Programme. It estimates that the ratio between top and bottom quintiles has narrowed slightly. Both estimates agree that nearer the extremes (top and bottom 10 percent and 5 percent), the ratio between the top and bottom continues to widen. In the global household survey, for example, World Bank economist Branko Milanovic estimates that the ratio between the top 5 percent was 114 to 1 in 1993, up from 78 to 1 in 1988 (Milanovic 2001; UNDP 2001) (see Tables 8.1 and 8.2).

Some other stunning comparisons emerged from the Milanovic survey:

- The richest 1 percent of people in the world receive as much income as the bottom 57 percent.

Figure 8.2   Distribution of World Income and Purchasing Power, 1993

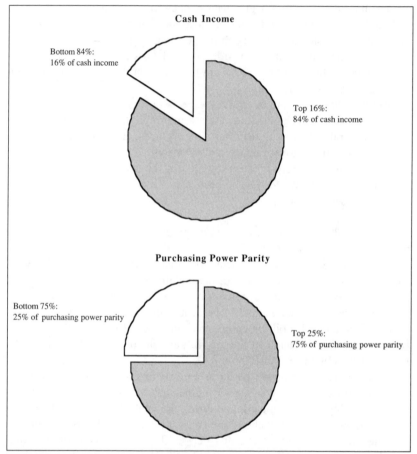

Source: Branko Milanovic, *True Income Distribution, 1988 and 1993* (Washington, DC: World Bank, 2001). Based on household surveys alone.

- A U.S. citizen having the average income of the bottom U.S. decile (tenth of the population) is better off than two-thirds of the world population.
- The top 10 percent of the U.S. population has an aggregate income equal to the income of the poorest 43 percent of people in the world (25 million people compared to nearly 2 billion.

Inequality in the United States. Using U.S. census data, in 1980 the richest one-fifth of U.S. households received 44.1 percent of the total income, while the poorest one-fifth had 4.2 percent—a ratio of 10.5. By 1998, the ratio

Table 8.1   Income Inequality Between the World's Richest and Poorest,
1970 and 1997 (based on country averages)

	Ratio Richest 10%: Poorest 10%		Ratio Richest 20%: Poorest 20%	
Measure	1970	1997	1970	1997
Exchange Rate	51.5	127.7	33.7	70.4
Purchasing Power Parity	19.4	26.9	14.9	13.1

Source: UN Development Programme, *Human Development Report, 2001* (New York: New York University Press, 2001). UNDP calculations based on World Bank data.

Table 8.2   Worldwide Real Per Capita Income by Percentile of Income
Distribution, 1988 and 1993

Percentile of Income Distribution	(1) Income in 1988	(2) Income in 1993	Ratio 2:1 (percentage)
5	277.4	238.1	86
10	348.3	318.1	91
15	417.5	372.9	89
20	486.1	432.1	89
25	558.3	495.8	89
30	633.2	586.0	93
35	714.5	657.7	92
40	802.7	741.9	92
45	908.3	883.2	97
50	1,047.5	1,044.1	100
55	1,314.4	1,164.9	89
60	1,522.7	1,505.0	99
65	1,898.9	1,856.8	98
70	2,698.5	2,326.8	86
75	3,597.0	3,005.6	84
80	4,370.0	4,508.1	103
85	5,998.9	6,563.3	109
90	8,044.0	9,109.8	113
95	11,518.4	13,240.7	115
99	20,773.2	24,447.1	118

Source: UN Development Programme, *Human Development Report, 2001* (New York: New York Univeristy Press, 2001). UNDP calculations are based on World Bank data.
Notes: All values expressed as purchasing power parity in 1993 international dollars. The values show income exactly at the given percentile of income.

had widened; the top one-fifth had 49.2 percent, the bottom one-fifth only 3.6 percent, and the ratio had widened to 13.7. Parallel to worldwide estimates, the poverty gap is widening even faster between the very top and bottom. The poorest are growing even poorer. The rich are getting much richer.

The World Bank uses per capita income instead of household income to compare income shares. They found the ratio between the rich and poor quintiles' incomes in the United States to be 8.9 in 1985, the highest among industrial nations. Among other wealthy nations, the ratio ranged from 3.4 and 3.6 in Denmark and Finland up to 6.5 and 7.0 for Australia and the United Kingdom (compare with the ratios in Table 8.3).

Inequality in developing countries. Among the low-income countries with estimates available, the ratios between the top and bottom quintiles ranged from 4.9 for Bangladesh and 5.7 for India up to 24.2 for Brazil (World Bank 2001a).

Differences in income distribution make a big difference to poor people. Brazil's per capita GDP is three times India's, but the poorest 20 percent of the population in each country has about the same purchasing power. Egypt's GNP is less than half Chile's, but Egypt's poor people have slightly greater purchasing power. Thailand and Colombia have approximately the same GNP, but Thai poor have more than twice as much purchasing power (see Table 8.3).

■ Direct Measures of Well-Being

Other indicators measure well-being even more directly than income or poverty rates—for example, infant or under-five mortality rates, life expectancy, educational achievement, and food intake.

Table 8.3   Poverty Impact of Income Distribution, Selected Countries, 1999

Country	GNP PPP$/Capita 1999	GNP-PPP$ 1999		Ratio Highest 20%: Lowest 20%
		Lowest 20%	Highest 20%	
South Korea	15,530	5,824	30,516	5.2
Chile	8,410	1,430	26,071	18.2
Malaysia	7,640	1,681	20,743	12.3
Brazil	6,840	889	21,546	24.2
Thailand	5,950	1,904	14,340	7.5
Colombia	5,580	837	16,991	20.3
Philippines	3,990	1,077	10,434	9.7
China	3,550	1,047	8,272	7.9
Egypt	3,460	1,695	6,747	4.0
Indonesia	2,660	1,197	5,466	4.6
India	2,230	903	5,140	5.7
Bangladesh	1,530	528	3,312	4.9
Nigeria	770	169	2,144	12.6

Source: World Bank, *2001 Development Indicators* (New York: Oxford University Press, 2001).

Hunger. The Food and Agriculture Organization of the United Nations (FAO) attempts to measure and estimate shortfalls in food consumption. In their *Sixth World Food Survey,* released in late 1996, the FAO estimated that the absolute number of people in developing countries who consumed too little food declined slightly over a period of two decades—from about 900 million in 1969–1971 to about 800 million in 1990–1992. Because population increased rapidly over the period, however, the proportion of people hungry in developing countries declined from about 35 percent to about 20 percent.

The most dramatic gains in reducing hunger over the period were in East and Southeast Asia, where the percentage of hungry people dropped from 41 to 16 percent and the number by nearly half—from 476 million to 269 million. Less dramatic gains by both measures were recorded in the Middle East and North Africa. The proportion declined, but the absolute number increased slightly over the period in South Asia and Latin America and the Caribbean. In Africa, the percentage of hungry people increased from 38 to 43 percent, while the number soared from 103 to 215 million (FAO 1996) (see Figure 8.3). Since 1992 the number of hungry people in developing countries has remained about the same, but the proportion has dropped to 19 percent (FAO 2001) (see Figure 8.4).

Human Development Index. The United Nations Development Programme has developed a Human Development Index (HDI), which gives equal weight to three factors: life expectancy at birth, educational attainment (based on the adult literacy rate and mean years of schooling), and per capita purchasing power (UNDP 2001).

People's lives can be improved if even limited resources are focused on nutritional programs, public health, and basic education. Both China and Sri Lanka, for example, have invested relatively heavily in education and health care since independence. They rank with many industrial countries in life expectancy and educational attainment.

Most of the formerly communist countries invested heavily in education and health care. Some former colonies continue to build on the educational systems established during the colonial era: Vietnam, Laos, Cambodia, and Madagascar (colonized by France); Guyana, Tanzania, Uganda, and Burma (colonized by Britain); and the Philippines (colonized by Spain and the United States). Several Latin American countries have emphasized education more recently: Chile, Costa Rica, Colombia, and Uruguay. In each instance, these countries rank higher on the HDI scale than other nations with a comparable per capita gross domestic product.

But sustaining such improvements requires steady or improving economic performance. Many of these nations have suffered recent economic downturns or are in the midst of drastic political and economic change. In

Figure 8.3 Distribution of Undernourished People, by Developing Region,
          1969–1971 and 1990–1992

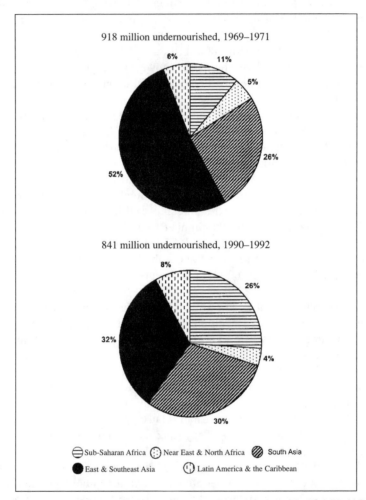

Source: Food and Agriculture Organization of the United Nations, *The Sixth World Food Survey* (Rome: FAO, 1996). Reprinted with permission.

the short term, at least, they are hard pressed to maintain their education and health programs.

Other nations rank much lower on the HDI scale than on a per capita GDP scale. The Middle Eastern oil-rich nations rank low in both longevity and educational attainment, particularly because of the status of women in

Figure 8.4 Number of Undernourished People, by Region, 1996–1998
(millions of people)

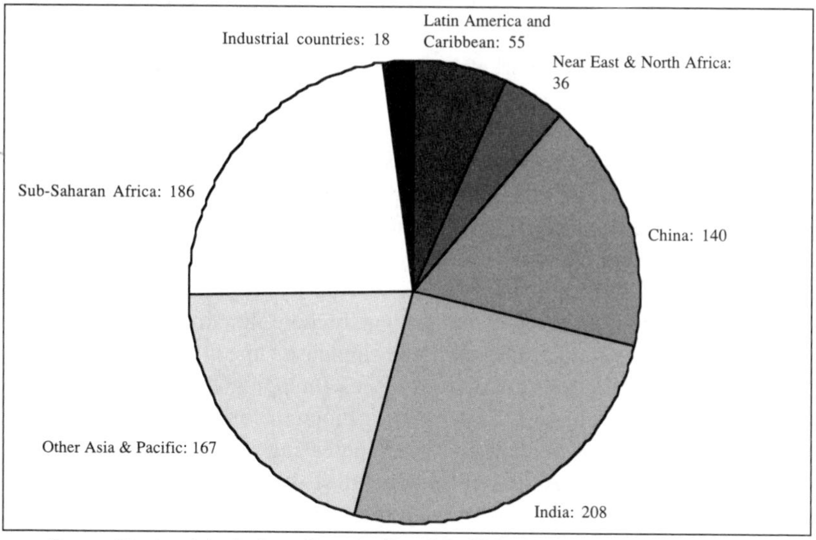

Industrial countries: 18

Latin America and
Caribbean: 55

Near East & North Africa:
36

Sub-Saharan Africa: 186

China: 140

Other Asia & Pacific: 167

India: 208

Source: Food and Agriculture Organization of the United Nations (FAO), *The State of Food Insecurity Around the World* (Rome: FAO, 2001).

these societies. Several African nations have extremely low educational attainment and longevity indicators, for varied reasons. Angola and Namibia have been engulfed in long independence struggles and civil war. Botswana and Gabon, although relatively rich in natural resource income, have not devoted proportional resources to education and health care services.

* * *

After this extensive digression to explore how poverty is measured, I turn now to the questions of why poverty is worse in some places than others and what might be done to reduce or eliminate it.

■ ECONOMIC GROWTH AND POVERTY REDUCTION IN A GLOBAL, KNOWLEDGE-BASED, MARKET ECONOMY

Individual, community, and national efforts to reduce poverty must be set in the context of the threefold revolution during the past two decades that has transformed national markets into a truly global economy:

- the evolution of a single worldwide system of producing and exchanging money, goods, and services;
- the shift from a resource-based economy to a knowledge-based economy; and
- the acceptance of market-based economics as conventional wisdom by most political leaders throughout the world.

The first two aspects of the global revolution are inextricably linked. New information, communication, and transportation technologies have dramatically changed the way many older businesses are managed (on-time delivery of manufacturing components, digital voice transcription, and automatic businesses-to-business [B2B] transmission of orders and billing, for example). These same technologies have also spawned whole new industries (distance learning and online marketplaces). Their immense capacity to process information has enabled new scientific advances (mapping the human genome, medical and space research, cross-species bioengineering). The knowledge factor outweighs the resource factor in an increasing number of endeavors.

At their peak in 1970, about 40 percent of the world lived in countries with centrally planned economies. By the turn of the twenty-first century, with a couple of minor exceptions, these countries had all introduced market-oriented reforms. Some, such as China, are introducing such reforms gradually. Others, such as the Soviet Union, held on until their economies collapsed. Despite rhetoric, no nation is attempting a true free market economy. Debates about political economies are usually about deciding which functions can be left to markets and which cannot, or shaping the context or rules under which markets function. A central question is how market-oriented political economies might contribute more to reducing poverty, especially in the poorest countries.

■ The Global Workforce: Need for 2 Billion Jobs

The route out of poverty for most people is through new economic opportunities—jobs or business ventures. As the world's population grows by one-third, from approximately 6.0 billion to about 8.0 billion by the year 2025, the global labor force will grow even faster, by about half, from 2.7 billion to more than 4 billion workers. In addition, the International Labour Organization estimates that more than a billion workers are now unemployed or underemployed (ILO 1996). Half of the new workers have already been born, and the number of unemployed is growing annually. The pressing need, therefore, is to create nearly 2 billion new economic opportunities during the next twenty-five years. Most of the new jobs or businesses will be needed in developing countries, where 95 percent of the increase in population and labor force is taking place.

Virtually all of the added jobs will need to be nonfarm. Governments in developing countries, or markets, may increase incentives for food production, but farmers are likely to adopt technologies that increase their productivity and reduce farm employment even faster. More and more farmers, or their children, will seek nonfarm employment. Whether such nonfarm employment is urban or rural will depend on policy choices. Improvements in education, health care, and public infrastructure can provide some public service jobs. But most new income-earning opportunities, if they come to pass, will be in the private sector.

Every new opportunity, whether public or private, for employee or self-employment, requires savings and investment—in human resources and in creating each job or business opportunity. The rates of savings and their allocation are crucial factors in determining whether enough decent income-earning opportunities can be created; these factors are determined in large measure by public policies.

■ Economic Growth and Poverty Reduction

Economic growth is often held up as the primary goal for economic development and as the means to increased employment opportunities. Some analysts, bankers, and political leaders equate *development* with *economic growth*. Most of these people expect poverty and other social problems to shrink as economies grow.

Economic growth is a necessary, but not sufficient, condition for reducing poverty. The distribution of the added income is also critical. Poverty has fallen rapidly in some fast-growing economies (Korea, Indonesia, China), while not changing much in others (Brazil, South Africa, Oman).

Because poverty is experienced in households and by individuals, detailed and accurate information at that level is critical. The World Bank has issued two recent studies of household surveys, covering 120 time comparisons, mostly during the early 1990s, in fifty developing economies (Ravallion and Chen 1997; Ravallion 2001). The surveys showed the following:

- Poverty rates have consistently fallen as average incomes have risen, and risen as average incomes have fallen.
- Poverty rates have not declined anywhere in the absence of economic growth.
- In developing countries, inequality increased as often as it decreased as average incomes rose.
- In transition economies (former communist states in Eastern Europe and Central Asia), inequality consistently increased at the same time as average incomes fell.

Other recent studies of eight East and Southeast Asian countries show that it is possible to have both economic growth and decreasing inequality if the right policies are in place. In South Korea, for example, where per capita income has grown rapidly, the most affluent fifth of the population has about six times as much income as the poorest fifth. The ratio has narrowed slightly over the past two decades; poor people have shared in the rapid growth. That is, their incomes have increased a little more than average on a percentage basis, although the absolute increase in income has been greater for wealthier persons.

In sharp contrast, Brazil's per capita GNP was twice Korea's in 1970. Since then, its economy has grown about half as fast; by 1999, Korea's per capita GNP was more than Brazil's, as measured by PPP. Although the income ratio between Brazil's poorest and richest fifth has narrowed somewhat, it remains about twenty-four fold. Poor Brazilians have scarcely benefited from growth and remain mired in deep poverty (see Table 8.3).

The Asian countries have reduced, or at least have not increased, economic inequality by giving poor people the incentive and the means to improve their own earning power. Examples are land reform and support for small farmers in Korea and Taiwan; high school education, especially for women, in Singapore; and manufacturing for export, which has raised the demand for unskilled factory workers, plus a massive affirmative action program for the poorer ethnic groups, in Malaysia.

■ The Virtuous Circle

Declining inequality and economic growth support each other in three ways, in an ascending, or *virtuous*, cycle:

- As poor families' incomes increase, they invest more in *human capital*—more education and better health care for their own children.
- Improved health and better education, which usually accompany decreased inequality, increase the productivity of poorer workers and their communities and nations. This in turn further increases their income.
- Greater equality contributes to political stability, which is essential for continued economic progress (see Figure 8.5).

Recent research shows that relative equality in distribution of national incomes increases the likelihood that economic growth can be sustained. Widespread participation in political as well as economic activity reduces the likelihood of enacting bad policies and permits their earlier correction (Birdsall, Pinckney, and Sabot 1996).

Figure 8.5   A Virtuous Circle

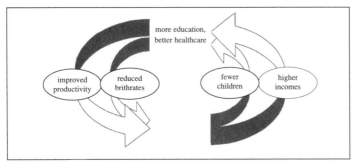

Source: Adapted from Bread for the World (BFW) Institute, *Hunger 1995: Causes of Hunger* (Washington, DC: BFW Institute, 1994). Used with permission of the BFW Institute.

◼ *Sustainable Development*

More than economic growth that is relatively equally distributed is needed to achieve long-term poverty reduction. The growth and distribution must be sustainable. The concept of *sustainable development* emerged in the 1970s to help ensure that development would not ruin the environment, which would, sooner or later, slow or reverse development. The UN World Commission on Environment and Development has defined *sustainable development* as "development which meets the needs of the present without compromising the ability of future generations to meet their own needs" (WCED 1987). More recently, and as used here, the term has evolved to incorporate other aspects of development in addition to economic growth. Advocates of international justice, environmental protection, peace, sustainable population growth, democracy, and human rights have increasingly come to see that their goals are interlinked. Several examples follow:

- There is no way to save the rain forests of Brazil without dealing with the need for land of poor Brazilians.
- There is no way to reduce rapid population growth in developing countries without improving living standards, especially for girls (see Chapters 9 and 10).
- There are no durable solutions to poverty and hunger in the United States without social peace, broader democratic participation, and a shift to economic patterns that will be environmentally sustainable.

Bread for the World is a nongovernmental organization (NGO) focused on policy decisions that would reduce hunger and poverty. It defines sustainable development in terms of four interconnected objectives: providing economic opportunity for poor people, meeting basic human needs, ensuring environmental protection, and enabling democratic participation (BFW Institute 1995). These concepts, and those from other chapters in this book, are reflected in the policy suggestions of the next section.

■ ANTIPOVERTY POLICIES IN A GLOBAL ECONOMY

Public policies aimed at reducing poverty fall into two broad categories: (1) creating appropriate, effective guidelines for markets and (2) collecting and allocating public resources, especially for investment in human resources.

The reality of achieving effective antipoverty policies is, of course, much more difficult than the assertion. Just as some actors in the marketplace can take advantage of their economic power, they and other powerful political actors can sway policies to their own self-interest, whether at the local, national, or international level. Meanwhile, poor people, whose well-being is the strongest evidence of whether policies are effective, often lack political access or clout. In addition, they and their allies are often unclear or divided on issues of national and international economic policy.

But as the global, knowledge-based, market economy reaches into the far corners of our planet, people of goodwill have only one option: to help draft and implement policies that will direct a sizable portion of this economy toward creating income-earning opportunities for poor people. The most important areas for policies to help reduce poverty include:

- investing in people;
- sustaining agriculture and food production;
- creating a framework for sustainable development; and
- targeting international financing (BFW Institute 1997; World Bank 2001b).

■ Investing in People

Health care and nutrition. Investments in basic health care and improved nutrition yield huge dividends. Healthy children learn better. Healthy adults work better. Improved health care begins with greater attention to basic public health measures: nutrition education, clean water and adequate sanitation, vaccination against infectious diseases, prevention of AIDS, distribution of iodine and Vitamin A capsules, and simple techniques of home health care. Delivery of these services can be relatively inexpensive, especially in developing countries, where village women with minimal training can be employed. These basic services should have priority over urban hospitals and specialized medical training.

In some instances, public health training can be delivered in conjunction with supplemental feeding programs such as the Special Supplemental Food Program for Women, Infants, and Children in the United States or the Integrated Child Development Services in India.

Education. Investments in basic education complement those in health care and improved nutrition and yield huge payoffs in both developing and industrialized nations. Better education for youth, especially girls, leads to improved health awareness and practices for their families on a lifelong basis. Cognitive and other skills improve productivity, enable better management of resources, and permit access to new technologies. They also enhance participation in democracy.

A study of ninety-eight countries for the period from 1960 to 1985 showed GDP gains up to 20 percent resulting from increases in elementary school enrollment and up to 40 percent resulting from increases in secondary enrollment. In allocating educational resources, the highest payoff is for elementary education, because it reaches the most children (Fiske 1993).

In the United States, dramatic improvement has followed investments in Head Start, which provides preschool education and meals for low-income children, and Job Corps, which provides remedial and vocational training for disadvantaged youth (see Figure 8.5).

■ Sustaining Agriculture and Food Production

Access to land. Widespread land ownership by small farmers usually contributes directly to food security and improved environmental practices. The more successful land reform programs, as in South Korea and Taiwan, have provided at least minimum compensation to existing landlords.

Equitable prices for farm produce. Thriving agriculture is basic to successful development in most of the poorest nations. Much of new savings must be accumulated within agriculture, since it is such a large share of the economy. Such savings are important for increasing agricultural productivity and for helping to finance rural, nonfarm businesses. Also, as their incomes rise, farmers expand their purchases of consumer goods, providing an important source of nonfarm employment.

In many developing countries, state-run marketing boards have taxed agriculture by setting farm prices very low and retaining for the government a large share of the value from farm exports. Meanwhile, the United States and the European Union have supported their own farmers in ways that generate surplus crops. They also subsidize exports of these crops, driving down prices around the world. Developing country farmers, who usually are not subsidized, cannot match the low prices. Agriculture falters and with it the whole process of development. Both rich country export subsidies and developing country discrimination against agriculture should be phased out as quickly as possible.

■ Creating a Framework for Sustainable Development

Access to credit. Equitable access to credit for small farmers and small businesses is probably the highest priority for the allocation of domestic savings or outside investment. Training in resource and business management is often part of successful credit programs.

Most informal economic activity results from the efforts of small entrepreneurs who cannot find a place in the formal economy. If they have access to good roads, markets, and credit, small farmers and small business people can create their own new income-earning opportunities in market economies.

Adequate physical infrastructure. Creating and maintaining an adequate physical infrastructure is essential to a viable, expanding economy. Important for rural areas are farm-to-market roads and food storage, both oriented to domestic production and, if appropriate, exports. For all areas, safe water, sanitation, electricity, and communications networks are needed.

Stable legal and institutional framework. Sustainable development requires a stable legal framework. This includes assured property titles, enforceable contracts, equitable access to courts and administrative bodies, and accessible information networks.

Stable currency and fiscal policies. Neither domestic nor international investors, including small farmers and microentrepreneurs, are likely to invest in countries in which the political or economic environment is unsettled. High inflation or continuing trade deficits, which often go together, discourage needed investments and may even drive out domestic savings.

Twenty-nine African countries undertook structural adjustment programs during the 1980s. These reforms have often fallen heavily on poor people, as social service programs were frozen or cut back to help balance national government budgets. Other measures to reduce budget or trade deficits—more progressive taxes, cuts in military spending, and curtailing luxury imports—have not always been pursued with equal vigor. Economic reform programs, which will continue to be necessary, should be revamped to reduce the costs to, and increase benefits for, poor people.

In 1999, led by the World Bank and the International Monetary Fund, global creditors undertook a new round of forgiving part of the debt of the poorest nations. The primary condition for debt reduction was that the savings would be invested in human services—primarily education and health care. This represented a significant change from the conditionality in earlier structural adjustment programs. In a second major change, plans to use the savings, as well as plans for new loans, were to be open for widespread comment by citizens and civil society groups. First results are promising.

Effective, progressive tax systems. Effective progressive tax structures are key to sustainable financing for investments in human resources and infrastructure. Taxes based on ability to pay are also key to stabilizing or reducing wide disparities in income distribution in both rich and poor countries. Such tax systems are usually difficult to enact where wealth and political power are controlled by a small minority.

Incentives for job-creating investments. The Asian countries that have grown so rapidly have all placed emphasis on labor-intensive exports—some in joint ventures with overseas partners, some with investments solicited from abroad, but many with subsidies from within their own economies. This is a distinct departure from their more general commitment to follow market signals.

A primary target for job-intensive investments will be processing operations for primary products—whether for domestic consumption or for crops and minerals now being exported. The success of these efforts will depend in considerable measure on further development of trade among poor nations and, especially, whether rich nations are willing to reduce higher tariffs on manufactured goods.

Targeting International Financing

The importance of international financing is well covered in Chapter 7. The only addition here is to emphasize the distinction between overall growth and growth that will create more and better opportunities for poor people.

SUMMARY AND CONCLUSION

Poverty—the lack of resources or income to command basic necessities—is the condition of about one-fifth of the world's population, or more than one-quarter of the people in developing countries. The absolute number of poor people has remained about steady in recent years, while their proportion has declined slightly, with considerable regional variation.

Countries need economic growth to overcome poverty, but other conditions are also critical. Relatively egalitarian distribution of national income among and within households matters greatly. Gains must be sustainable. Decisionmaking must be broadly shared.

Creating 2 billion good jobs or business opportunities is the biggest single challenge for this generation. The economic and policy tools to generate relatively equitable growth have been successfully demonstrated in recent years, particularly in East and Southeast Asia.

Meanwhile, some of the worst effects of poverty have been, and should continue to be, offset by public and private interventions: Infant mortality and overall hunger have declined, and literacy and longevity have increased in many instances, even in the face of continued poverty.

Adapting these tools and programs to particular circumstances, especially in Africa, is of utmost concern to everyone. In an increasingly global economy, the well-being and security of each person or community or nation is inescapably linked to that of every other.

Finally, overcoming poverty requires the one ingredient that seems in shortest supply: the political will to do so.

■ DISCUSSION QUESTIONS

1. Are you more inclined to measure poverty in terms of absolute income, income distribution, or the capacity to reach more fundamental goals? If the last, what would be your list of goals?
2. Which of the antipoverty policies do you think are the most beneficial?
3. What would you consider a reasonable goal for the ratio between the top and bottom income groups within an economy? Within a business firm? What policies would be necessary to move toward these goals?
4. Should government policies encourage the redistribution of income? If so, to what extent?
5. Are you as optimistic as the author that poverty can be overcome?
6. Do you concur that the well-being of everyone is inescapably linked?
7. Do you agree with the author that the principal missing ingredient in overcoming hunger is political will?

■ NOTE

This chapter is adapted from an earlier work by the author in *Hunger 1995: Causes of Hunger* (BFW 1994) and is used with permission of Bread for the World Institute.

■ SUGGESTED READINGS

Birdsall, Nancy, Thomas Pinckney, and Richard Sabot (1996) "Why Low Inequality Spurs Growth: Savings and Investment by the Poor," Inter-American Development Bank Working Paper Series, No. 327. Washington, DC: IDB.

Birdsall, Nancy, David Ross, and Richard Sabot (1995) "Inequality and Growth Reconsidered: Lessons from East Asia," *World Bank Economic Review* 9, no. 3.

Bread for the World Institute. Annual Report on the State of World Hunger. Washington, DC: BFW Institute.

Drêze, Jean, and Amartya Sen (1989) *Hunger and Public Action.* Oxford: Clarendon Press.

FAO (Food and Agriculture Organization of the United Nations) (last published in 1996) Occasional World Food Survey. Rome: FAO.

——— (2001) *The State of Food Insecurity in the World.* Biennial. Rome: FAO.

Fiske, Edward B. (1993) *Basic Education: Building Block for Global Development.* Washington, DC: Academy for Educational Development.

United Nations Development Programme (annual) *Human Development Report.* New York: Oxford University Press.

World Bank (biennial) *World Development Indicators.* New York: Oxford University Press.

——— (annual) *World Development Report.* New York: Oxford University Press.

Part 3

DEVELOPMENT

9

POPULATION
AND MIGRATION

Ellen Percy Kraly

Coming to grips with the implications of current population trends is an extremely important dimension of global studies. The process is neither easy nor comforting, because significant population increase is an inevitable characteristic of the global landscape in the first fifty years of the twenty-first century. It is critical that students interested in global issues should appreciate both the causes of population growth and the consequences of population change for society and the environment. Such an appreciation will serve in developing appropriate and effective responses to population-related problems emerging globally, regionally, and locally.

This chapter seeks to contribute to the understanding of the interconnections among population change; environmental issues; and social, economic, and political change in both developing and developed regions of the world. Because population growth has momentum that cannot be quickly changed, it is important to begin by considering fundamental principles of population or demographic analysis and to place recent global and regional population trends in historical perspective. This chapter examines the widely divergent philosophical and scientific perspectives on the relationships among population, society, and environment that have pervaded visions of the future. Debates on the implications of current growth have also influenced discussions about routes for population policy. In the next section, global effects of population redistribution, urbanization, and international migration are discussed. The chapter concludes by considering global dimensions of population policies targeting growth and international population movements.

■ PRINCIPLES AND TRENDS

▦ Demographic Concepts and Analysis

Demography is the study of population change and characteristics. A population can change in size and composition as a result of the interplay of three demographic processes: fertility, mortality, and migration. These components of change constitute the following population equation:

P = (+) births (–) deaths (+) in-migration (–) out-migration;
where P is population change between two points in time

On the global level, the world's population grows as the result of the relative balance between births and deaths, often called natural increase. The U.S. population is currently increasing at about 0.9 percent per year; natural increase accounts for about two-thirds, and net international migration constitutes about one-third of this relatively low level of population growth.

Many people seeking routes to sustainable development advocate a cessation of population growth often referred to as zero population growth (ZPG). When viewed from a short-run perspective, ZPG means simply balancing the components of the population equation to yield no (zero) change in population size during a period of time. Population scientists, however, usually consider ZPG in a long-term perspective by considering a particular form of a zero-growth population: A stationary population is one in which constant patterns of childbearing interact with constant mortality and migration to yield a population changing by zero percent per year. In such a case, fertility is considered replacement fertility because one generation of parents is just replacing itself in the next generation. In low-mortality countries, replacement-level fertility can be measured by the total fertility rate and is approximately 2.1 births per woman to achieve a stationary population over the long run.

It takes a relatively long time, perhaps three generations after replacement fertility has been achieved, for a population to cease growing on a yearly basis. Large groups of persons of childbearing age, reflecting earlier eras of high fertility, result in large numbers of births even with replacement-level fertility. Hence, an excess of births over deaths occurs until these "age structure" effects work themselves out of the population. This is known as the momentum of population growth. To illustrate, Wolfgang Lutz estimated that if the world's populations achieved replacement-level fertility instantaneously, global population would have continued to grow from 5.3 billion in 1990 to 7.4 billion in 2050, an increase of 40 percent (Lutz 1994: 57).

Age structure is an important social demographic characteristic of a population. Both the very young and the very old in a population must be supported by persons in the working age groups. The proportions of persons in different age groups in a population are depicted in a *population pyramid*. Figure 9.1 shows population pyramids for two countries, Tanzania and Spain. The pyramid for Tanzania reveals a youthful population with nearly one-third of the population under ten years of age. This reflects high fertility. In Spain, a low-fertility country, only 9 percent of the population is under ten years. At the other end of the age spectrum, less than 5 percent of the Tanzanian population is over the age of sixty years, compared to over one-fifth, 21 percent, of the Spanish population. The immigration of male workers into Spain is also evident in the large proportion of males aged thirty to thirty-four. These two age pyramids illustrate the history of past levels in fertility and migration as well as the different demands on society for support for the young and the old.

■ Historical and Contemporary Trends in Population Growth

The world's population was estimated to be 6.1 billion in mid-2000 and increasing at approximately 1.2 percent per year (UNPD 2001b). These data represent a cross-sectional perspective on population characteristics—a snapshot that fails to capture the varying pace of population change worldwide and regionally. Over most of human history, populations have increased insignificantly or at very low annual rates of growth, with local populations being checked by disease, war, and unstable food supplies. Between the sixteenth and the eighteenth centuries, population growth appeared to become more sustained as a result of changes in the social and economic environment: improved sanitation, more consistent food distribution, improved personal hygiene and clothing, political stability, and the like.

The world's population probably did not reach its first billion until just past 1800. But accelerating population growth during the nineteenth century dramatically reduced the length of time by which the next billion was added. According to the United Nations (UNDESIPA 1994; UNPD 2001b), world population reached

- 1 billion in 1804,
- 2 billion in 1927 (123 years later),
- 3 billion in 1960 (33 years later),
- 4 billion in 1974 (14 years later),
- 5 billion in 1987 (13 years later),
- 6 billion in 1999 (12 years later).

Figure 9.1   Ages of Males and Females as Percentage of Population, Spain and Tanzania, 2000

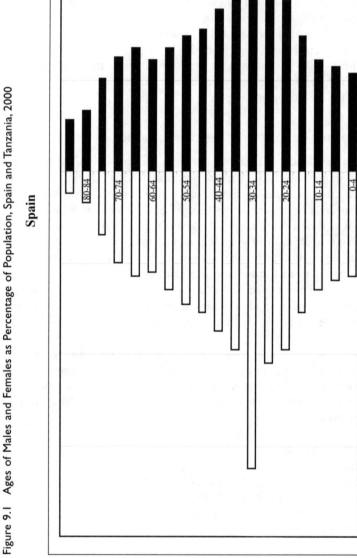

Spain

Figure 9.1   continued

Tanzania

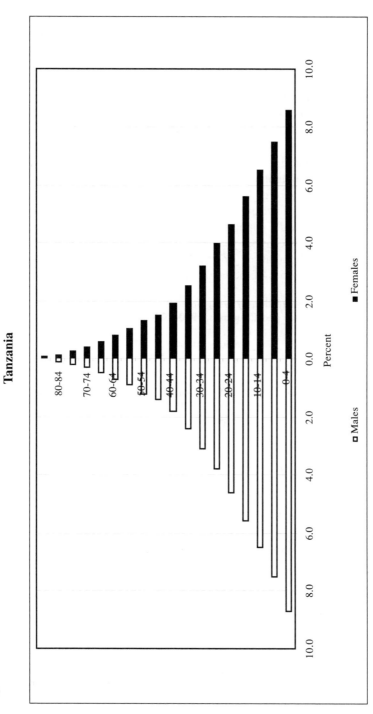

Source: UN Population Divison, *World Population Prospects: The 1998 Revision.* Vol. 2, *Sex and Age* (New York: United Nations, 1999).

Rapid population growth occurred on a global scale during the second half of the twentieth century. Population data for years since 1950 are shown in Figure 9.2. Between 1950 and 2000, the world's total population increased from 2.6 billion to 6.1 billion.

The difference in height of the bars in Figure 9.2 reveals the momentum of population growth that results in continued additions to the world's population, albeit in decreasing numbers: between 1985 and 1990, approximately 85 million persons were added to the world's population each year; in the late 1990s, the annual increase is estimated at 77 million (UNPD 2001b).

It is important to note, however, that in spite of these large additions to the world's population, the rate of population growth is *decreasing*. The average annual rate of global population growth reached an all-time high of about 2.2 percent between 1962 and 1964. Since that time, the pace of the growth of the world's population has decreased to the current rate of approximately 1.2 percent per year.

Patterns of population growth differ significantly between more and less developed regions of the world. Table 9.1 provides greater geographic

Table 9.1 World Population by Geographic Region and for More and Less Developed Countries, 1950–2000

	1950	1960	1970	1980	1990	2000
Population (in millions)						
World	2,521	3,022	3,696	4,440	5,266	5,772
More developed countries	813	916	1,008	1,083	1,148	4,601
Less developed countries	1,709	2,106	2,688	3,358	4,119	1,171
Africa	221	277	357	467	615	732
Asia	1,402	1,702	2,147	2,641	3,181	3,428
Latin America and the Caribbean	167	218	285	361	440	489
Europe	547	605	656	693	722	800
North America	172	204	232	255	282	295
Oceania	13	16	19	23	26	29
Geographic region (percent distribution)						
World	100.0	100.0	100.0	100.0	100.0	100.0
More developed countries	32.2	30.3	27.3	24.4	21.8	79.7
Less developed countries	67.8	69.7	72.7	75.6	78.2	20.3
Africa	8.8	9.2	9.7	10.5	11.7	12.7
Asia	55.6	56.3	58.1	59.5	60.4	59.4
Latin America and the Caribbean	6.6	7.2	7.7	8.1	8.4	8.5
Europe	21.7	20.0	17.7	15.6	13.7	13.9
North America	6.8	6.8	6.3	5.7	5.4	5.1
Oceania	0.5	0.5	0.5	0.5	0.5	0.5

Sources: UN Population Division, *World Population Prospects: The 2000 Revision* (New York: United Nations, 2001); UN Population Division, *World Population Prospects: The 1998 Revision.* Vol. 1, *Comprehensive Tables* (New York: United Nations, 1999).

Figure 9.2 World Population for More and Less Developed Countries, 1950–2000

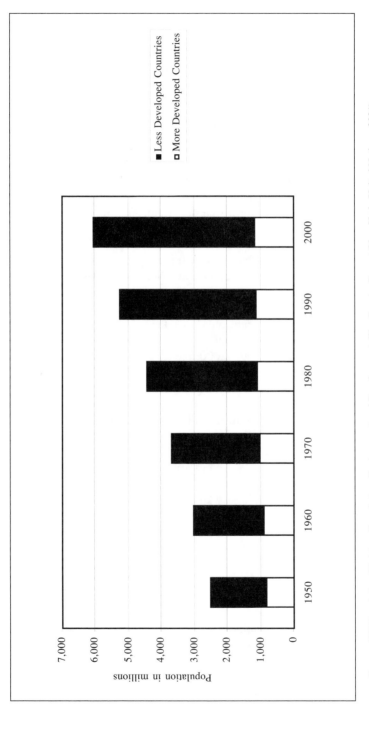

Source: UN Population Division, *Population, Environment and Development: The Concise Report* (New York: United Nations, 2001).

detail and summarizes population size and distribution for major regions of the world for selected years since 1950.

Dramatic shifts in the geography of world population have occurred during the past five decades and, as discussed below, are expected to continue well into the future. In 1950, just over two-thirds of the world's population was located in less developed countries; by 2000, this proportion had increased to four-fifths. Asian countries make up three-fifths of the world's population; over one-fifth, 21 percent, of the global village lives in China; and another 17 percent live in India (data not shown in Table 9.1). Africa's share has increased from just under 10 percent in 1950 to about 13 percent of the current population. European populations constituted 13.9 percent of the world's population in 2000, a decline from 22 percent in 1950. Western Hemisphere regions—North America, Latin America, and the Caribbean—include approximately 14 percent of the world's population.

Population growth is fueled by levels of fertility, mortality, and net migration. The rapid population growth that occurred in the post–World War II era reflected significant declines in mortality resulting in large part from public health advances and the transfer of medical technology from more to less developed countries.

The total fertility rate measures the average number of births per woman of childbearing age and is a strong indicator of overall population growth. In the period 1995–2000, the total fertility rate for the world as a whole was 2.8 births per woman, representing a significant decline from 4.2 in about 1985. Fertility in more developed countries has been below replacement for some time and is currently estimated at 1.6 births per woman. In developing countries, the rate has dropped from 4.7 in 1985 to 3.1 in 1995–2000. Much of this decline is weighted by the aggressive fertility control campaign in China and significant declines in fertility throughout Southeast Asia and in Latin America. Total fertility in India has also declined but less dramatically, from 4.3 in 1985 to 3.3 currently. Total fertility in Africa, particularly sub-Saharan Africa, while declining somewhat in the past decade, remains strikingly high: 5.3 births per woman currently. Figure 9.3 provides a cartographic view of current levels of fertility for countries of the world.

■ Perspectives on the Causes and Consequences of Population Change

Reflections on the relationship between population and society can be found in the early history of many cultures. In early Greece, Plato wrote about the need for balance between the size of the city and its resource base; Confucianism emphasized the social and economic advantages of large families. Concern about the implications of population growth for

Figure 9.3   World Fertility Rates (number of births per woman), 1995–2000

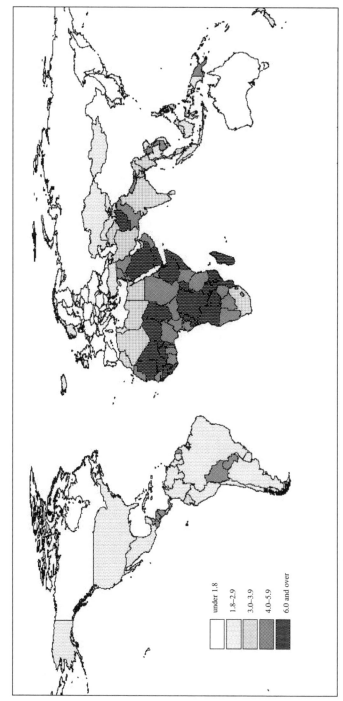

under 1.8

1.8–2.9

3.0–3.9

4.0–5.9

6.0 and over

Source: UN Population Division, *World Population Prospects: The 2000 Revision* (New York: United Nations, 2001).

social progress became a focus of social theory in the nineteenth century and continues in contemporary debates on the global effects of current levels of population growth.

■ Debates on Population Growth

Certainly the most influential statement concerning the sources and implications of population growth was that of Thomas Malthus in his *Essay on the Principle of Population;* the first edition, published in 1798, was followed by several revisions (Malthus 1826). In his essay, Malthus was reacting to mercantilist philosophy that pervaded eighteenth-century European thought and emphasized the value of large and increasing populations for economic growth and prosperity. Malthus offered a negative perspective on the consequences of population growth, arguing that increases in population will outpace increases in food supply, resulting in starvation, poverty, and human misery. Because of the instinct of humans to reproduce, population growth can only be halted through the so-called positive checks— rising mortality as a result of famine, war, and epidemics. Malthus, a clergyman, opposed contraception and advocated delayed marriage and abstinence, which he called "preventive" checks. Neo-Malthusians, in contrast, recognize the importance of birth control as a means to limit family size and hence reduce population growth.

Among the early critics of Malthusian thought was Karl Marx, who argued that there exists no universal law of population like the one Malthus had generated; rather that consequences of population growth derive from the particular form of economic organization within a society. According to Marx, capitalism, not population growth, resulted in increasing levels of poverty. These historical perspectives on population hold a true vitality for contemporary debates concerning population issues and policies.

Two lines of thought have dominated the population debate in the past three decades and might be crudely labeled *neo-Malthusian* and *cornucopian.* The neo-Malthusian perspective continues to emphasize the problem of population growth as the primary obstacle to sustaining the ecological balance of planet earth by leading to natural resource depletion, pollution, and loss of biodiversity. Cornucopian perspectives, on the other hand, emphasize the role of technological innovation and market forces, which through pricing effects will manage the use of natural resources. From this vantage point, population growth holds potential for solving global problems through increased economic productivity and capacity for technological progress.

Because of the dramatically different visions of society and nature inherent in these two perspectives, the debate between the negative and positive consequences of population growth has always been energetic, animated, and often contentious. It is therefore helpful to many concerned students of population studies that a third perspective on population, society, and environment linkages has emerged in the past few decades. This perspective has been labeled by some as *structuralist* (see Harper 1995).

Structuralist perspectives, which borrow from Marxist theory, consider population characteristics, including population growth, poverty, food supplies, and environmental problems, as outcomes of broader social structural processes and institutions. Thus, population growth, specifically high fertility, is more a consequence than a cause of slow economic development and restricted social and economic opportunities. Barry Commoner views poverty and the low status of women as factors contributing to high fertility in developing societies; more significant for the quality of the global environment, moreover, are the high levels of energy consumption and waste in rich Northern countries (Commoner 1992).

From these disparate perspectives, Charles Harper (1995) sees an "emerging consensus" concerning the consequences of population growth. This perspective conceptualizes population growth as both cause (like the neo-Malthusians) and consequence (like the structuralists) of social, economic, and environmental processes confronting the world, nations, and local communities. Hence, the reduction of poverty, the improvement of life chances and of the status of women, the increasing sustainability in food production, the improvement in water quality, and the like, all become important strategies for reducing population growth. Also recognized is the advantage of slowing and ultimately ceasing world and regional population growth in order to more effectively improve standards of living, stabilize food supplies, and halt environmental degradation (see UNPD 2001a and National Research Council 1986).

Helping to chart the progress toward low population growth in societies is the model of the demographic transition that was developed initially as a description of population growth patterns in Europe and North America in the nineteenth and early twentieth centuries. The model became linked to theories of modernization to predict how population growth would proceed throughout the developing world. As societies underwent industrialization and urbanization, with all the concomitant social changes, death rates would fall as they had in the Western societies, followed with a lag by declines in fertility. During the lag, population growth would occur until norms and values concerning the need for large families were replaced by small family ideals.

The demographic transition model has been widely criticized, retested, and revised. A major limitation of the theory is the view that the experiences of non-Western societies will mirror or converge with those of Europe and North America. Revisions of the demographic transition model present a clearer understanding of how family size is influenced by cultural beliefs and by gender, particularly educational opportunities for girls.

Moreover, studies in sub-Saharan Africa and South Asia have shown the importance of understanding the contributions of children to the well-being of the family and its kin. Young children in developing societies may be a source of wealth as household labor and family prestige, and older children may provide old-age support and continuation of the family lineage (Caldwell 1982).

■ EXPECTATIONS ABOUT FUTURE POPULATION GROWTH

Theories of population change guide analyses of future population growth, usually in the form of population projections. Most demographers are quick to state that population projections are not predictions but rather represent a calculation of future population size based on a set of assumptions or variants. Shown in Figure 9.4 are population estimates and projections prepared by the United Nations Population Division for the years 1950–2050. The world totals reflect the sum of projections conducted for 228 countries or areas; the "fan" of population figures represents the four projection variants for the projection period 1990–2050. These variants reflect different assumptions about the pace and pattern of fertility change.

The constant fertility variant assumes that fertility levels for 1995–2000 are maintained throughout the projection periods. In the high variant, fertility is assumed to reach a total fertility rate of 2.6. In the low variant, fertility is assumed to reach 1.7 births per woman. The medium variant targets fertility at approximately replacement level, 2.1 births per woman. In all four projections, mortality levels decline, although at a slower pace in those countries where mortality is already low. Hence, the projection series as a whole assumes convergence among countries to common fertility levels and mortality levels, generally embodying a demographic transition model. The high-fertility scenario results in growth in world population from 6.1 to over 12 billion between 2000 and 2050, an increase of nearly 5 billion. The low scenario projects an increase to 7.9 billion in 2050, an increase of nearly 30 percent. Thus, even with convergence of fertility levels to well below replacement (a global total fertility rate of 1.7), the world's population will continue to grow as a result of the momentum of population growth. The medium variant, which models trends toward replacement fertility, results in an increase to 9.3 billion in 2050, an increase of 50 percent over the 2000 population.

It is important to note that in contrast to previous projections of future population growth prepared by the United Nations, the revised projections for 2000 reveal the impact of the HIV/AIDS epidemic. In the forty-five countries most affected by the disease, it is estimated that 15.5 million deaths due to HIV/AIDS will occur between 2000 and 2050. In spite of the high toll of the epidemic, however, high levels of fertility in these countries

Figure 9.4   Projected World Population to 2050 (by fertility variants)

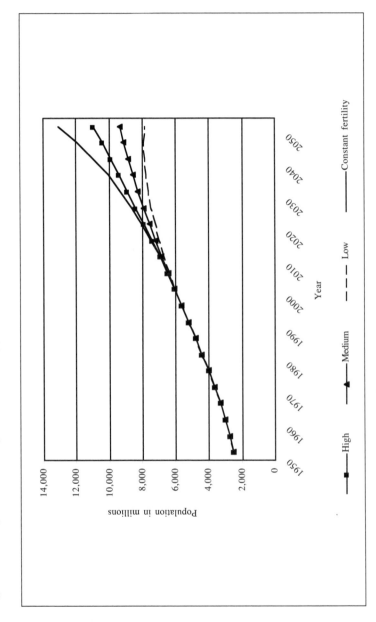

Source: UN Population Division, *World Population Prospects: The 2000 Revision* (New York: United Nations, 2001).

will continue to result in overall population growth. For example, in Botswana, Swaziland, and Zimbabwe, countries with infection rates of over 25 percent, population is projected to increase by 37, 148, and 86 percent, respectively, between 2000 and 2050. (UNPD 2001b: vii).

Shifts in the geographic distribution of the world's population are evident in all of the projections, as seen in Table 9.2. African countries, primarily in sub-Saharan regions, will increase from 13 percent to 22 percent of the world's population, and the population of Europe is expected to decline from 12 percent to 7 percent of the world's population. Countries in North America and Latin America and the Caribbean will continue to constitute about 14 percent of the world's population. Asian countries will continue to hold the largest share of the world's population, although decreasing from 61 percent to approximately 58 percent. It is important to consider two Asian countries in particular, China and India, whose population policies are considered later in this chapter. The population of China is projected to

Table 9.2 Projections of World Population by Geographic Region and for More and Less Developed Countries, 2000 and 2050

| | | Fertility Assumptions | | | |
| | | Low | Medium | High | Constant |
	2000	2050	2050	2050	2050
Population (in millions)					
World	6,057	7,866	9,322	10,934	13,049
More developed countries	1,191	1,075	1,181	11,309	1,162
Less developed countries	4,865	6,791	8,141	9,625	11,887
Africa	794	1,694	2,000	2,320	3,566
Asia	3,672	4,527	5,428	6,430	7,376
Latin America and the Caribbean	519	657	806	975	1,025
Europe	727	556	603	654	580
North America	314	389	438	502	446
Oceania	31	42	47	53	56
Geographic region (percent distribution)					
World	100.0	100.0	100.0	100.0	100.0
More developed countries	19.7	13.7	12.7	103.4	8.9
Less developed countries	80.3	86.3	87.3	88.0	91.1
Africa	13.1	21.5	21.5	21.2	27.3
Asia	60.6	57.6	58.2	58.8	56.5
Latin America and the Caribbean	8.6	8.4	8.6	8.9	7.9
Europe	12.0	7.1	6.5	6.0	4.4
North America	5.2	4.9	4.7	4.6	3.4
Oceania	0.5	0.5	0.5	0.5	0.4

Source: UN Population Divison, World Population Prospects: The 2000 Revision (New York: United Nations, 2001).

increase from 1.3 billion in 2000 to 1.5 billion in 2050; India's population is projected to grow from 1 billion in 2000 to 1.6 billion in 2050, thus overtaking China as the largest national population in the world. Thus, we can expect significant shifts in world geography of population based on these projections.

■ SOCIAL AND ENVIRONMENTAL DIMENSIONS OF POPULATION REDISTRIBUTION AND MIGRATION

As well as influencing population change, the movement of persons within and among countries is both a cause and a consequence of social, economic, political, and environmental factors. Geographic mobility is the general concept covering all types of human population movements. Migration is generally considered to refer to moves that are permanent or longer term; internal migration within a country is distinguished from international population movements; international migration to a country is *immigration,* and international migration from a country is *emigration.* Reasons for moving are often included in migration concepts, such as labor migration, refugee migration, and seasonal migration. Internal and international migration are processes that are increasingly linked through the geographic and social dimensions of global economic development.

Internal Migration and Urbanization

A corollary of the demographic transition model is growth in the size of cities as well as increasing proportions of populations living in cities and metropolitan areas, that is, urbanization. The UN estimates that 47 percent of the world's population was living in urban areas in 2000; differences between more and less developed countries are dramatic, 76 percent and 40 percent, respectively. It is important to remember that while some of the largest metropolitan areas in the world are found in developing countries—for example, Mexico City and Bombay each with 18.1 million in 2000, São Paulo with 17.8 million, and Calcutta and Shanghai each with 12.9 million—the majority of populations in these societies currently live in rural areas (UNPD 2001c).

Given population growth and rural-to-urban migration, however, urbanization is expected to increase throughout regions of the world. By the year 2030, for example, the level of urbanization in more developed countries is expected to increase to 84 percent and in less developed countries even more steeply to 56 percent, a dramatic shift in patterns of residence and economic activity (UNPD 2001c).

The causes of urban growth have varied among regions and during different historical periods. In Western societies, urbanization was fueled in large part by technological change in both agricultural and industrial sectors, resulting in both a push from rural communities and a pull to emerging industrial centers (Harper 1995). In developing societies, rural-to-urban migration is driven by many factors, including increasing population density (caused by high fertility rates) in rural areas, environmental degradation from practices such as overgrazing, and the hope for gainful employment in urban areas. The pull of cities in many developed countries exists in the form of the hope for employment and higher wages rather than prearranged jobs. As a result, levels of unemployment in cities in developing countries are very high. Evidence of underemployment is shown in the large numbers of persons, including many children, attempting to earn livelihoods in what has been called by some the informal economy—for example, street vendors, curbside entertainers, and newspaper boys and girls.

■ International Population Movements

One of the most visible manifestations of globalization is the increasing scale of international population movements throughout all regions of the world. According to scholars Stephen Castles and Mark Miller,

> Millions of people are seeking work, a new home or simply a safe place to live outside their countries of birth. For many less developed countries, emigration is one aspect of the social crisis which accompanies integration into the world market and modernization. . . . Migrations are not an isolated phenomenon: movements of commodities and capital almost always give rise to movements of people. Global cultural interchange, facilitated by improved transport and the proliferation of print and electronic media, also leads to migration. . . . Migration seems likely to go on growing into the new millennium, and may be one of the most important factors in global change. (Castles and Miller 1998: 3–4)

The consequences of international population movements for both sending and receiving nations and communities will have significant implications for emerging global issues.

Countries in both Western and Eastern Europe have been faced with large numbers of persons seeking political asylum from both European regions as well as geographically distant sources, including East Africa and Southeast Asia. Significant labor migration flows have emerged between South and Southeast Asia and oil-producing regions of the Middle East and throughout the Asia Pacific region. The United States has grappled with issues concerning large numbers of undocumented migrants drawing from Mexico and many other source countries. Refugee migration is recurringly characteristic of political and environmental change throughout Africa.

Castles and Miller identify five "general tendencies" of contemporary international population movements that they expect to continue well into the twenty-first century. First, international population movements will involve an increasingly large number of countries, both as sending and receiving regions, hence the *globalization of migration*. Second, the volume of international migration can be expected to *increase* in volume. Third, international migration will continue to become *differentiated* by including a wider variety of migrants, for example, seasonal migrants as well as migrants seeking permanent resettlement. Fourth, as women throughout the world become increasingly involved in the global workforce, international migration will become more *feminized*. Fifth, international migration is likely to become a more significant *political issue,* both on the international stage as well in the politics of individual nations (Castles and Miller 1998: 8–9).

The significance of the scale of refugee migration and displaced populations in global population issues cannot be overstated. Refugees seek safety from war and oppression but can also be a source of political and economic instability in border regions and countries of asylum. For instance, Afghanis seeking refuge following the U.S. military response to the September 11 terrorist attacks in 2001 faced resistance in neighboring Pakistan.

The international definition of a refugee is a person who

> owing to a well-founded fear of being persecuted for reasons of race, religion, nationality, membership of particular social group(s) or political opinion is outside the country of his nationality and is unable to or owing to such fear is unwilling to avail himself of the protection of that country; or who, not having a nationality and being outside the country of his former habitual residence . . . is unable or unwilling to return to it. (UNHCR 1995: 256)

At the end of 2000, the Office of the UN High Commissioner for Refugees (UNHCR) identified more than 21.1 million persons throughout the world who were of concern to the organization. Of this extraordinary number, 12.1 million are recognized as refugees living in asylum in other countries; the remainder are persons who are internally displaced within their own countries for complex political, economic, and environmental reasons and persons outside their home country in refugee-like situations. Table 9.3 displays refugees by broad category and geographic region estimated by the UNHCR for 2000. Many refugee settlements or camps have existed for many years. Some refugees have been repatriated to their homelands, for example, Guatemalans who had sought refuge in Mexico and Muslims who had fled Myanmar (formerly Burma); others, including many Vietnamese during the 1970s and 1980s, have been permanently resettled in other countries such as Canada, Australia, and the United States. The majority of the world's refugees are women and children, whose voices are

Table 9.3 Refugees and Other Types of Persons of Concern to the UNHCR,
by Geographic Region, 2000

	Total	Refugees	Asylum Seekers	Returned Refugees	Internally Displaced and Others of Concern
Persons of concern (in thousands)					
World	21,794	12,072	914	786	8,022
Africa	6,073	3,626	87	271	2,089
Asia	8,450	5,383	46	351	2,670
Latin America and the Caribbean	576	38	3	1	534
Europe	5,571	2,325	354	164	2,728
North America	1,048	631	417	0	0
Oceania	75	68	7	0	0
Geographic region (percent distribution)					
World	100.0	100.0	100.0	100.0	100.0
Africa	27.9	30.0	9.5	34.5	26.0
Asia	38.8	44.6	5.0	44.7	33.3
Latin America and the Caribbean	2.6	0.3	0.3	0.1	6.7
Europe	25.6	19.3	38.7	20.7	34.0
North America	4.8	5.2	45.7	0.0	0.0
Oceania	0.3	0.6	0.8	0.0	0.0

Source: UN High Commissioner for Refugees, Population Data Unit, *Refugees by Numbers, 2000 Edition* (New York: Oxford University Press, 2001).

often not heard in discussions about programs to aid and resettle refugees (UNHCR 2001).

■ POPULATION POLICIES

A direct or explicit population policy is a "strategy for achieving a particular pattern of population change" (Weeks 1999: 510). According to the UN survey of government policies, seventy-five countries have policies or programs to reduce population growth; twenty countries, including Cyprus, Greece, Hungary, and Gabon, have goals to increase growth (UNPD 2000). Many more countries have indirect population policies that, while not targeting population growth, have clear implications for either mortality, fertility, or migration. The United States, for example, has not yet adopted a formal statement of goals concerning national population growth but does have a long-standing policy for the permanent resettlement of immigrants and refugees (see discussion below), which in turn results in net additions to the population through international migration.

Source: Joel Pett, *Lexington* (Kentucky) *Herald Leader,* July–August 1995.

International Efforts to Reduce Population Growth

International population conferences bring government delegations and representatives of nongovernmental organizations (NGOs) together to discuss goals concerning population and to develop strategies for achieving those goals. The most recent conference was held in Cairo in 1994. In each of these gatherings, there has been a general recognition, though not universal agreement, that (1) rapid population growth fueled by high fertility poses a challenge to economic development in less developed countries; (2) mortality should be reduced regardless of the effect on population growth; and (3) international migration is an appropriate arena of national policy and control (Weeks 1996). Over the past three decades, however, important shifts in thinking about population growth have occurred that have had implications for the agenda of population policies and programs within countries. These are discussed in the next section.

The 1974 World Population Conference in Bucharest produced the first formal expression of a world population policy. The "World Population Plan of Action," however, embodied a wide range of perspectives on the ways to reduce population growth within developing societies. Some countries, notably the United States, advocated fertility control, specifically family planning programs, to reach population growth targets. Other countries, primarily

in the developing world, emphasized the role of development in leading to
fertility decline (hence, "development as the best contraceptive"). The 1984
International Population Conference in Mexico City found the United
States reducing its support for family planning (which was linked in turn
to the Reagan administration's views on abortion) and identifying population growth as having little hindrance on economic and social development.
Many developing countries by this time, however, had instituted family
planning programs in an effort to slow the retarding effects of rapid population growth on improving standards of living and educational levels and
reducing mortality (Weeks 1996).

The 1994 International Conference on Population and Development
recognized the global dimensions of population change. The conference report, "Programme of Action," identifies the connections among population
processes, economic and social development, human rights and opportunities, and the environment, thus shifting attention away from targets concerning population growth to goals concerning sustained development, reduction of poverty, and environmental balance. The central role of women
in the goals and programs to achieve sustainable development is underscored in the final report: "The key to this new approach is empowering
women and providing them with more choices through expanded access to
education and health services, skill development and employment, and
through their full involvement in policy- and decision-making processes at
all levels" (UNDESIPA 1995: 1).

The emphasis on connecting population issues to the status of women
represents a significant forward step in embedding population analysis into
broader discussions of the quality of human life and the balance between
society and environment at local, national, and global levels.

Over many decades, the U.S. government has implemented programs
abroad concerning population, development, and human welfare. As mentioned earlier, U.S. support for international family planning programs has
wavered depending on the current president's political lens. Under President Ronald Reagan, support for fertility control programs was barred from
those countries that permitted women access to legal abortion. This restriction on U.S. foreign aid was lifted in the early days of the Clinton administration, only to be reinstated under President George W. Bush.

■ National Population Policies

Most countries that have formal population policies seek to reduce population growth by reducing fertility. Beginning in the 1950s, providing contraception through family planning programs was initiated in many developing countries—often with significant contributions from developed
countries, NGOs and foundations, and, in subsequent decades, international

organizations such as the UN Fund for Population Activities (UNFPA). Increasingly, policies aimed at fertility reduction have encompassed broader perspectives on population dynamics, incorporating goals to increase the status of women through better health, enhanced educational and employment opportunities, greater access to credit, and so on.

The record of family planning programs has been variable throughout the developing world. India has maintained a national family planning program since 1952 that has met with fertility trends that vary significantly throughout regions within the country and between rural and urban areas. In 1973, Mexico instituted a national policy to reduce population growth, with focus on the reduction of fertility through maternal and child health programs, family planning services, sex education, and population information programs.

China's fertility control policy began in 1971 as a set of policy goals (*wan xi shao*) concerning later marriage (*wan*), longer intervals between births (*xi*), and fewer children (*shao*), with the one-child policy implemented in 1979. During the 1970s the Chinese birthrate declined significantly reflecting the provision of contraception in combination with social and economic incentives to delay and reduce fertility. The birth control program also coincided with a general decline in fertility, which had been evident since the early 1960s. The Chinese fertility policy has been criticized harshly, however, for being coercive and for leading to selective abortion, abandonment, and infanticide of girl infants. The current level of fertility in China is estimated to be 1.8 births per woman.

The demographic implications of rapid fertility decline in China can be anticipated using population projections. In spite of below-replacement fertility the Chinese population is projected to increase from 1.275 billion in 2000 to 1.462 in 2050 (medium-fertility variant). This 15 percent increase in population size is accompanied with significant shifts in age composition: in 2000, 10 percent of the Chinese population was over sixty years of age; by 2050, this proportion is expected to triple to 30 percent. (By way of contrast, the proportion of the U.S. population currently over sixty years is 16 percent.) This dramatic aging of Chinese society in the near future will require changes in Chinese social and economic arrangements to support and care for an increasingly elderly population (UNPD 2001b).

▪ International Migration and Refugee Policies

The UN Declaration of Human Rights recognizes the basic right of people to leave their homelands. The converse of this right to emigrate, however, is not recognized—that is, nation-states have the sovereign right to control the entry of nonnationals into their territory. Nearly all countries have clear policies concerning international migration and travel. The very few countries that continue to allow international migration for permanent resettlement include the United States, Canada, Australia, and New Zealand, often considered the traditional immigrant-receiving countries. A much larger range of countries provide humanitarian assistance to refugees in the form of programs admitting refugees for permanent resettlement, response to requests for political asylum, land for refugee camps, and provision of financial and other resources for international organizations seeking to respond to refugee situations.

The demand for international migration, temporary labor migration, refugee resettlement, and temporary asylum can be expected to continue to grow in the next decade, with those seeking to move drawing overwhelmingly from developing regions. Emerging and persistent patterns of undocumented migration throughout both the developed and the developing world are symptoms of the motivation of people to seek better opportunities through international migration. Matched with this demand are national doors that are gradually closing to international migrants. In the traditional receiving countries, concerns over the social, economic, political, and security effects of immigration have moved high on the political agenda at both the national and state or provincial levels, often, but not always, with efforts to tighten migration controls, reduce immigration levels, and constrain access of immigrants to national and local social programs.

The United States accepts the largest number of immigrants for permanent resettlement in the world. Since the mid-1990s immigration for permanent residence to the United States has averaged approximately 800,000 persons per year. In 1998, Mexico, China, India, and the Philippines provided the largest numbers of immigrants to the United States.

U.S. immigration policy is organized around several principles. First, the policy gives priority to close relatives of U.S. citizens and to a lesser extent relatives of immigrants already in the country. Second, the policy gives priority to persons with occupations, skills, and capital that will benefit the U.S. economy. This dimension of U.S. immigration policy may contribute to the loss of highly skilled and professionally trained persons from developing countries (referred to as "brain drain"). Third, immigrants to the United States must be admissible on the basis of a long list of personal characteristics (for example, good health, lack of criminal background, and sufficient economic resources). Fourth, there are annual numerical limits on the major categories of immigration to the United States.

Refugees are resettled in the United States if they meet criteria of the international definition provided above. The numbers of persons admitted as refugees reflects international need identified by the U.S. State Department in consultation with Congress. In the mid-1990s, an average of 80,000 refugees were admitted annually, largely from Bosnia-Herzegovina, countries in the former Soviet Union, and Vietnam (USDJ 2000). Since the

September 11 attacks, refugee admissions to the United States have been sharply curtailed.

The United States also issues hundreds of thousands of "nonimmigrant" visas each year to persons visiting the country for specific purposes such as tourism and business, university study, consulting, and temporary employment. One likely effect of the September 11 terrorist attacks in New York and Washington, D.C., will be more restriction of temporary visas and generally greater scrutiny of all visa applications. The relationship between international migration and homeland security has become a significant dimension of the U.S. immigration debate.

■ CONCLUSION

Population trends and patterns within countries and regions hold fundamental and inescapable implications for the full spectrum of global issues addressed in this book. While the annual rate of world population growth has been declining in recent decades, significant increases in population size, particularly in countries in the developing world, will continue into the near future. Understanding the sources of population change, specifically declines in fertility and patterns of migration, is a critical dimension of efforts to attain sustainable development and reduce poverty in regions throughout the world and to protect the global environment.

To contribute to these important discussions, this chapter has sought to introduce readers to the international and national population issues that exist and are emerging within the global arena. The discussion has been built on a foundation of basic tools for the study of population, important trends in population growth and its component demographic processes, and dominant perspectives concerning the causes and consequences of population growth. Threaded throughout the discussion has been consideration of the relationships among perspectives about population, society, and environment; models of future population growth; and the design of global and national population policies.

■ DISCUSSION QUESTIONS

1. Is population growth a major global problem?
2. Do you agree more with the views of the cornucopians or the neo-Malthusians with reference to population growth?
3. Should countries open their borders to refugees fleeing political persecution and those seeking economic opportunity?

4. Was the Chinese government's one-child policy justified? Should governments be involved in population policy?
5. Should the U.S. government give foreign aid to reduce world population growth? Should the aid be conditional?
6. To what groups of immigrants should countries give preference?

■ SUGGESTED READINGS

Arizpe, Lourdes M., Priscilla Stone, and David C. Major, eds. (1994) *Population and Environment: Rethinking the Debate*. Boulder, CO: Westview Press.

Ashford, Lori S. (1995) "New Perspectives on Population: Lessons from Cairo," *Population Bulletin* 50, no. 2. Washington, DC: Population Reference Bureau.

Castles, Stephen, and Mark J. Miller (1998) *The Age of Migration: International Population in the Modern World*, Second edition. New York: Guilford Press.

Commoner, Barry (1992) *Making Peace with the Planet*. New York: New Press.

Harper, Charles L. (1995) *Environment and Society: Human Perspectives on Environmental Issues*. Upper Saddle River, NJ: Prentice Hall.

McFalls, Joseph Jr. (1995) "Population: A Lively Introduction," *Population Bulletin* 46, no. 2. Washington, DC: Population Reference Bureau.

Moffett, George D. (1994) *Critical Masses: The Global Population Challenge*. New York: Penguin.

Myers, Norman, and Julian L. Simon (1992) *Scarcity or Abundance: A Debate on the Environment*. New York: W. W. Norton.

UNHCR (United Nations High Commissioner for Refugees), Population Data Unit (2001) *Refugees by Numbers, 2000 Edition*. Available at http://www.unhcr.org/.

UNPD (United Nations Population Division) (1999a) *World Population Prospects: The 1998 Revision*. Vol. 1, *Comprehensive Tables*. New York: United Nations.

——— (1999b) *World Population Prospects: The 1998 Revision*. Vol. 2, *Sex and Age*. New York: United Nations.

——— (2000) *Global Population Policy: Database, 1999*. New York: United Nations.

——— (2001a) *Population, Environment and Development: The Concise Report*. New York: United Nations.

——— (2001b) *World Population Prospects: The 2000 Revision*. New York: United Nations.

——— (2001c) *World Urbanization Prospects: The 1999 Revision*. New York: United Nations.

U.S. Department of Justice (2000) *1998 Statistical Yearbook of the Immigration and Naturalization Service*. Washington, DC: Government Printing Office.

Weeks, John (1999) *Population: An Introduction to Concepts and Issues*. Belmont, CA: Wadsworth.

10

WOMEN AND DEVELOPMENT

Elise Boulding with Jennifer Dye

This chapter uses the term *development* to refer to social, economic, and political structures and processes that enable all members of a society to share in opportunities for education, employment, civic participation, and social and cultural fulfillment as human beings, in the context of a fair distribution of the society's resources among all its citizenry. The United Nations (UN) bound itself in its Charter "to achieve international cooperation in solving international problems of an economic, social, cultural, or humanitarian character." In other words, the more industrialized countries of the North agreed to help the less industrialized countries of the South to reach the higher economic and social level already achieved in the North. The first thing to note about this development planning is that it has been done almost entirely by men, for men, with women and children as a residual category. How could half the human race be invisible to development planners in spite of Mao Tse-tung's well-known saying that "women hold up half the sky"? This chapter will review how this situation evolved, how it has hampered "real development," and what women—and men—are doing about it.

■ FROM PARTNERSHIP TO PATRIARCHY

In the early days of the human species, men, women, and children moved about in small hunting and gathering bands, sharing the same terrain and exploring the same spaces. Role differentiation was minimal, though childbearing restricted women's movements somewhat, and men ranged farther in hunting prey. Even so, it is estimated that women, as roving gatherers, supplied up to 80 percent of a band's diet by weight and therefore carried an important part of society's ecological knowledge in their heads. Based on observation of small hunting-gathering tribes found in Africa, Australia, and South America today, it would seem that old women as well as old men took the role of tribal elders and carried out rituals important to the social life of the band. About 12,000 B.C.E., a combination of events brought about a major change in the human condition; improved hunting techniques resulted in a dwindling supply of animals, and women discovered from their plant-gathering activities that seeds spilled by chance near last year's campsite would sprout into wheat the following year, creating a convenient nearby source of food. This resulted in the deliberate planting of seeds to grow food, and agriculture came into being. This changed everything. Since working with seeds was women's work, women became the farmers, and men went farther and farther afield in search of scarce game, thus discovering exciting new terrain that women knew nothing about. The shared-experience worlds of women and men were now differentiated. Women knew about everything that was close to home, and men were exploring the world "out there."

While a lot has happened to humankind since 12,000 B.C.E., bearing children, growing and processing food, and feeding families (also creating shelters and clothing) have continued to be women's work for most of the world's population. Moving from the earliest farming settlements to villages to towns to cities and civilizations took another 8,000 to 10,000 years, but cities never returned to women that freedom of movement they had had as gatherers. Rather, they heralded the rise of the rule of men, or patriarchy, to replace an earlier partnership. What cities and civilization brought were concentrations of wealth and power—and houses that enclosed women and shut them off from the outside world. Only poor women were free to scurry about the streets to provide services for the rich. With industrialization, and major population movements to the cities where the factories were located, there were fewer and fewer women enjoying the relative freedom of farming and craft work, which families often carried out as a family team in the period known as the Middle Ages. Now they were either shut up in the home or shut up in factories.

While we have been describing major trends for women, reality for individual women was much more complex, and in every age women found ways to be creative and to improve the circumstances in which they lived. While most of women's creativity was invisible, we do find women—queens and saints and philosophers and poets—in the history books. We also find precursors of the contemporary women's movement from the 1500s through the 1700s. However, by the nineteenth century, the pressures of urbanization and industrialization in Europe and North America gave rise to a new social sector of educated middle-class women with free time and a growing awareness of the world around them, including awareness of women migrants from rural areas, chained with their children to factory and slum. This produced a small group of women radicals and revolutionaries and a larger group of liberal reformers and concerned traditionalists who translated their sense of family responsibility into responsibility for the community. These women quickly discovered, when they intruded into the political arena, that as women they had no civic or legal identity, no political rights, and no economic power. The realization that they needed civic rights to get on with reform led to an exciting century and a half of mobilization of women in the public arena. Their concerns about economic conditions for the poor soon spilled over to concern about the frequent wars that rolled over Europe, which they saw as directly related to poverty and suffering. The suffrage movement came into being because women realized that they needed political power in order to fight the social evils they saw. Everywhere, women came up against the rule of men and sought partnership with them rather than domination by them.

The patriarchal model pervades society but begins in the family with the rule of the male head of household over wife or wives and children; it has served as a template for all other social institutions, including education, economic life, civic and cultural life, and governance and defense of the state. Because so many generations of humans have been socialized into the patriarchal model, the struggle to replace it with partnership between women and men will be a long one; and it has barely begun. Now, however, the survival of the planet is at stake. Women's knowledge of their social and physical environment, of human needs, of how children learn, of how conflicts can be managed without violence and values protected without war—as well as their skill in managing households with scarce resources— are urgently needed wherever planning takes place and policies are made. Yet these are precisely the places where women are absent. Now we will turn to the story of how women are beginning to sow the seeds of transformation of a system based on domination over the very ones who hold the skills to shape a better life for all. First, we look at the United Nations and its role in this process.

■ THE UNITED NATIONS: THE DEVELOPMENT DECADES

As Western colonial empires began to dissolve in the UN's first decade in the 1950s, it became clear that the ever more numerous new, poor nations that came to be called the "third world" were going to need a lot of assistance from the founding states of the UN in order to work their way out of poverty. The popular "trickle-down" theory, which asserts that financial benefits given to big business and upper-class sectors will in turn pass

down to smaller businesses, consumers, and lower-class sectors, led to the
encouragement of capital-intensive heavy industry in poor countries. This
type of industry requires large expenditures of capital. The result is a loss
of labor-intensive light industries that require labor and might have employed the unemployed and the unschooled. The only aid given to agriculture was in the form of capital-intensive equipment that required large
acreage and pushed subsistence farmers onto ever more marginal lands with
rapidly deteriorating soils. This approach, labeled economic dualism, leaves
the bulk of a country's population without the skills and tools needed to be
more productive. By the end of the 1950s, it was evident that "international
cooperation" was not helping poorer countries at all. Then the UN General
Assembly declared the 1960s a Development Decade, a time of catching up
for all the societies left behind in the twentieth-century march of progress.
Yet the policies based on economic dualism remained unchanged, in spite of
valiant efforts by the UN Development Programme (UNDP), established in
1965, to undertake a more diversified approach. By the end of that decade,
gross national product (GNP) growth rates were negative in a number of
third world countries, even in some that had "good" development prospects.

A second Development Decade was launched in 1970 to try to do what
the first had failed to do, but by then it had become clear to many third
world countries that the development strategies recommended by "first
world" countries were not working. They formed their own Group of 77
(eventually including 118 states) to promote a Program of Action for the establishment of a New International Economic Order. This called for more
aid from countries of the North, debt moratoriums, and other strategies for
preventing the economic gap between the North and South from continuing
to increase. A monitoring system for the conduct of multinational corporations was also demanded. The North had no serious interest in acting on
these proposals, and the poor nations kept getting poorer. In 1980, the UN
launched yet another Development Decade, still following failed policies.
One problem faced was the World Bank's structural adjustment program,
which compelled countries of the South to focus on producing cash crops
for the international market to reduce their indebtedness, and to spend less
money on human services. This intensified the dualism between a low-pro-
ductivity agricultural sector and a high-productivity agribusiness and industrial sector. People in the South went hungry and without schooling
while food was exported to the North (since more money could be made),
and countries sank still deeper into poverty.

The goal of all of this Development Decade activity, planned and carried out by men, was to increase the productivity of the male worker. But
one quiet day in the late 1960s, a Danish woman economist, Ester Boserup,
decided to study what was actually going on in that intractable subsistence

sector of agriculture. She was able to point out that the majority of the food producers were women, not men, but all agricultural aid, including credit and tools, was given to men. This aid not only failed to reach the women but encouraged men to grow cash crops (which needed irrigation) for export. This added to the women's work load, as the men demanded that their wives tend the new fields in addition to the family plot that fed the family and provided a modest surplus for the local market.

Now a whole new picture began to emerge. Boserup focused her studies on Africa, which was bearing the cruelest load of suffering in the food supply crisis. She found that roughly 75 percent of food producers were women, often either sole heads of households or with migrant husbands living and working elsewhere. Their working hours were on the average fifteen hours a day. They worked with babies on their backs, and their only helpers were the children they bore. These women farmers never received advice on improved methods of growing food, tools to replace their digging sticks, wells to make water more available, credit to aid them during the years of bad crops, or aid in marketing. Long hours of walking to procure water and to get to market (with their produce carried on their heads), in addition to the hard work in the fields, made their lives a heavy struggle.

Yet emphasizing the hardship women experience should not obscure the importance of women's special knowledge stock, which is basic to community survival in any country, in both rural and urban settings. In a future world with more balanced partnering between women and men, knowledge will not be so gender linked; for now, it is imperative that it be recorded, assessed, and used. Women's special knowledge stock relates particularly to the six following areas: (1) life-span health maintenance, including care of children and the elderly; (2) food production, storage, and short- and long-term processing; (3) maintenance and use of water and fuel resources; (4) production of household equipment, often including housing construction; (5) maintenance of interhousehold barter systems; and (6) maintenance of kin networks and ceremonials for handling regularly recurring major family events as well as crisis situations. While only (2) relates directly to food, all six factors contribute to the adequate nutrition of a community; (6) is particularly important in ensuring food sharing over great distances in times of food shortages and famine. Unfortunately, the extended-kin and ceremonial complex is one of the first resources to be destroyed with modernization.

The story of the UN's Third Development Decade could have been different if policy planners had had this information at their fingertips. But now there was a women's movement ready to hear what Ester Boserup and a growing group of women development professionals had to say about women's roles in the development process. The UN came to play an important part in giving them a platform.

■ THE UNITED NATIONS: THE WOMEN'S DECADES

In 1963, the UN General Assembly proclaimed 1965 International Cooperation Year (ICY). Just two years before, the Women's Strike for Peace
movement had spread from the United States to other continents, so when
word got out that 1965 would be a UN International Cooperation Year, an
international women's network quickly sprang into being to organize projects for the year in many countries.

Between the women's inexperience and the international bureaucracy's
inertia, the ICY fell short of its sponsors' ambitious hopes; but for the international community of women, it proved to be the start of a still unfolding
chain of events. Women formed traveling teams to every continent and
began seeing with their own eyes in the rural areas of Africa, Asia, and
Latin America what Ester Boserup was documenting as an economist. They
began to see women as the workers of the world, though invisible to census
enumerators because much of their productive labor takes place in the
world of the home: the kitchen, the kitchen garden, and the nursery. "Employed persons" were supposed to have identifiable outside jobs and wages.
Unwaged labor (such as child care, housework, etc.) and labor in what is
called the informal economy (which is not reported in the economy's record
books, such as street vendors, prostitution, and illegal drug sales) were concepts not taken seriously by economists and statisticians.

This did not stop the women who traveled together in the ICY teams.
As they began to see homemakers as workers, they redefined their own
role. They liked to be called "housekeepers of the world." "International
housekeeping" brought them in 1972 to the UN Conference on the Human
Environment in Stockholm and in 1974 to the World Population Conference
in Bucharest and the World Food Conference in Rome. Each time, they
came in larger numbers and with better documentation on how the conference subject was relevant to women. They also became increasingly aware
of how blind most of their male colleagues were to the importance of
women's roles in economic production and social welfare. Since governments appointed few women to international conferences, the knowledge
and wisdom of this growing group of observers went unheard. They were
outsiders, petitioners, protesters.

Nevertheless, the UN had actually established, in its Economic and Social Council, a UN Commission on the Status of Women as early as 1946,
and by 1948 the UN Declaration on Human Rights included gender as a category. In fact, over the decades the UN adopted twenty-one conventions on
the rights of women, although they remained largely paper rights because
member states did not provide for their implementation. It was still a man's
world. But women were learning strategies for working in it. The Commission on the Status of Women in particular was building up experience in

working within the UN system and began to assert itself. Its members saw to it that the phrase "integrating women into development" was included in the development program for the Second Development Decade. While ignored by decisionmakers, the phrase immediately reverberated in the growing international women's movement. Furthermore, in 1972, Helvi Sipila of Finland was appointed the first woman UN assistant secretary-general, for humanitarian and social affairs. Empowered by Sipila's support, the Commission on the Status of Women worked closely with the older women's organizations and the newer networks to create a women's agenda for the UN.

The first real breakthrough came when the UN General Assembly declared that 1975 would be International Women's Year. The Women's World Plan of Action (UN 1976), drafted for the 1975 assembly of women in Mexico City, defined *status* in terms of the degree of control women had over their conditions of life. This became a key theme that continues to this day in the women's movement. The International Women's Year became the UN International Decade for Women (or Women's Decade), with follow-up meetings in 1980 in Copenhagen, in 1985 in Nairobi, and in 1995 in Beijing. These UN-sponsored world women's conferences have represented a growing voice for women as participants and coshapers of the world in which they live. The trio of themes for each of these gatherings, which swelled from 6,000 women in Mexico City to 14,000 women in Nairobi to 50,000 in Beijing, has been: equality, development, and peace. The interrelationship of these three concepts in the lives of women, and in the life of every society, represents an important breakthrough in the conceptualization of development.

The guidelines for national action laid down for the Women's Decade are as relevant today as they were twenty years ago:

- involving women in the strengthening of international security and peace through participation at all relevant levels in national, intergovernmental, and UN bodies;
- furthering the political participation of women in national societies at every level;
- strengthening educational and training programs for women;
- integrating women workers into the labor force of every country at every level, according to accepted international standards;
- distributing more equitably health and nutrition services to take account of the responsibilities of women everywhere for the health and feeding of their families;
- increasing governmental assistance for the family unit;
- involving women directly, as the primary producers of population, in the development of population programs and other programs

affecting the quality of life of individuals of all ages, in family
groups and outside them, including housing and social services of
every kind.

Immediate outcomes of the 1975 conference included the establishment of the UN International Research and Training Institute for the Advancement of Women (INSTRAW), the UN Development Fund for Women
(a voluntary fund), and a modest working relationship with the UN Development Programme. With few resources and very small staffs, the two new
UN bodies have nevertheless played an important part in the gradual recognition of women as actors, not only as subjects needing protection.

Since the UN declared the international Women's Decade, some promising steps have been taken within the international community toward progress
for women in development. One of the most notable and better known events
is the 1995 Fourth World Conference on Women in Beijing. Since this conference the United Nations has also called a special session, the Beijing +5,
designed to review and assess past and current progress as well as future actions and initiatives. In addition, in 1997 the UN's Division for the Advancement of Women (DAW), the United Nations Development Fund for
Women (UNIFEM), and INSTRAW founded WomenWatch. WomenWatch
was created to monitor the results of the 1995 Beijing Conference as well
as to create Internet space on global women's issues.

In short, women were no longer to be treated as invisible or as subject
to patriarchal rule. They are persons, citizens, and actors on the local and
global scene. The importance of the role of the UN in providing an international platform for women to be seen, heard, and listened to with respect
cannot be underestimated—even though the UN itself still has a long way
to go in making senior posts available to women. Now the concepts are
there, in public international discourse, and the UN and its member states
cannot completely ignore them, however much they drag their feet in ap-
plying them. This change in visibility could not have happened without a
strong involvement of women from every continent. How did the invisible
become visible so quickly?

■ WOMEN'S NETWORKS REDEFINE
THE MEANING OF DEVELOPMENT

As mentioned earlier, women were already active internationally on behalf
of the oppressed poor by the middle of the nineteenth century. They came
to know each other across continents by meeting at the great world's
fairs—London in 1851, Paris in 1855 and 1867, and Chicago in 1893.
These were women of the upper middle classes who could travel, of course.

Many innovations in education and welfare services and home services for working women and children, prisoners, and migrants emerged from those meetings. By 1930, there were thirty-one international nongovernmental women's organizations (INGOs or NGOs) and more of them had working-class members. However, it was not until the United Nations brought grass-roots women from each continent in touch with each other that there emerged a deeper understanding of the problems that women faced in countries formerly colonized by the West. While every country experienced gender-based dualism (women's-only and men's-only jobs), only in the third world was gender-based dualism linked to an economic dualism that trapped women in the subsistence sector. In other words, women were left without the skills or opportunity for economic improvement.

While the cultural specifics of the situation of women differed among countries and continents, they all had a common base of experience linked to the fact of bearing and rearing children and having responsibility for the nourishment, health, and well-being of family members. Thus, while men measured development in economic terms of rates of growth of the gross national product, women's thinking was in terms of human and social development. What made life better for individual human beings, and what made societies more humane and joyful to live in? These were the questions women were asking as they formed numbers of new women's INGOs committed to human rights, development, and an end to violence. The same questions were being asked by the new women's networks that began forming during preparations for the first International Women's Year.

The first of the new networks was formed at the International Tribunal on Crimes Against Women, which met in Brussels in 1974. From that tribunal we get the first major international statement about patriarchal power as violence against women per se, apart from specific abusive acts. Participants joined to establish the International Feminist Network after listening to a horrifying array of violence experienced by women from all classes on all continents at the hands of men in their families, communities, and places of work. Today we are aware that this violence extends to the horror of wartime rape, an old but increasingly visible crime against women wherever the violence of war takes place.

As awareness of the obstacles to achieving the good life not only for women but for society as a whole increased, new transitional networks developed and multiplied from the time of the tribunal to the present. Thus, internationally women are strongly aware that obstacles persist to their full partnership in the development process and that violence and other human rights abuses and war itself constitute a significant part of those obstacles.

Since being able to work with men as equals is a precondition for women's voices to be heard, equality has been a goal at least since the beginning of the suffrage movement. But gaining the vote, with some associated legal and property rights, did not improve women's economic situation, nor did it bring sexual discrimination to an end or move women toward policymaking positions in society. Therefore, one part of the equal rights movement has dealt in painstaking detail with legal rights in the area of family, employment, rights to education and training, and equal access to opportunity in general. This is important for women everywhere but especially in the poorest countries, where men's literacy rate is barely at 40 percent and women's literacy rate is only a third or a half of the men's. As opportunities for advanced study become more available to women, the increase in women graduating with degrees in law, engineering, and the sciences, both physical and social, means that the sisterhood for social change on each continent can draw on a high level of competence and expertise in its work for equality.

Another part of the equal rights movement has taken a more direct political track: Women have sought to be involved in the lawmaking process as elected officials. Women's representation in politics can be broken down into parliamentary and executive branches of government. Since 1987, women's representation in single and lower chambers of parliaments around the world has increased by 2 percent, starting at 9 percent in 1987, remaining at 9 percent in 1995, and growing to 11 percent in 1999. Women are still greatly underrepresented in national parliaments throughout the world. However, in certain regions, women's representation has made great strides, specifically in the Caribbean, Latin America, and Eastern and Southeastern Asia (see Table 10.1) (UNSD 2001).

In addition, women are underrepresented in executive branches of government (including the following positions: president or head of state and prime minister and ministry positions and cabinet) around the world. Since 1974, only seventeen countries have elected a woman president, and since 1960, women have been elected prime minister in only twenty-two countries. In 1994, women held no ministerial positions in fifty-four countries, but by 1998 this number had dropped to forty-five countries. Yet around the world there is evidence that women's representation is slowly increasing. In 1998, twenty-eight countries had women in 15 percent of ministerial positions, versus sixteen countries in 1994 that had women in 15 percent of ministerial positions.

Women's political participation may be thought of as a pyramid, with the bulk of women found at the bottom, at the local level, and the fewest at the top, at the diplomatic level. In general, there is a tipping-point phenomenon for women in elected or appointed office. When there are a very few, "token" women, they tend to confine themselves to what are thought of as traditional women's issues, notably the well-being of families and children. As women grow in numbers and self-confidence, however, they are empowered to address broader systemic issues that affect all sectors of society.

Table 10.1   Representation of Women in National Parliaments, by Region, 1987, 1995, and 1999

Region	1987	1995	1999
World average	9.0	9.0	11.0
Africa			
Northern Africa	3.0	4.0	3.0
Sub-Saharan Africa	7.0	9.0	10.0
Latin America and the Caribbean			
Caribbean	9.0	11.0	13.0
Central America	8.0	10.0	13.0
South America	7.0	9.0	13.0
Asia			
Eastern Asia	18.0	12.0	13.0
Southeastern Asia	10.0	9.0	12.0
Southern Asia	5.0	5.0	5.0
Central Asia	0.0	8.0	8.0
Western Asia	4.0	4.0	5.0
Oceania	2.0	2.0	3.0
Developed regions			
Eastern Europe	26.0	9.0	10.0
Western Europe	14.0	18.0	21.0
Other developed regions	7.0	12.0	18.0

Source: UN Statistical Division, *The World's Women 2000: Trends and Statistics* (New York: United Nations, 2001).

As professional skills are bringing women into the legislative process, their know-how is also becoming available to development specialists, including the language of *development alternatives* and what is termed *another development.* Instead of high-tech industrial enclaves and agribusinesses that rob the poor of their intensive hand labor, alternative development means investing in tools and water supplies for farmers and craft workers; making credit available for small improvements, including equipment for small local factories; and building local roads, schools, and community centers that can serve many needs, including day care for small children and health services.

While women have been leaders in all of these efforts to achieve human and social development, not simply economic development, they have certainly not been alone. Ever since E. F. Schumacher (1993) wrote about "economics as if peopled mattered," there have been creative and humanistically oriented male development professionals who have worked both alone and together with women to further these broader goals. They too have helped to strengthen the women's networks.

Protecting the environment from multinationals that are logging whole forests and leaving fragile soils to erode, that are undertaking mining operations destructive of local farmland and waterways, that are building dams that destroy huge acreage and cause an increase in homelessness and unemployment—these activities also become an important agenda for women's networks. They are proud to bear the label *ecofeminists*. (It was a woman, Ellen Swallow, who coined the term *ecology* in 1892, and it was a woman, Rachel Carson, who is credited with initiating the birth of the modern environmental movement with her book *Silent Spring,* appearing in 1962.)

■ FROM PATRIARCHY TO PARTNERSHIP

We are midstream in a long, slow process of transformation. As described more fully at the end of *The Underside of History: A View of Women Through Time* (Boulding 1992), in one sense the institutions of society continue to be stacked against women. There are strong expectations of subservient behavior on the part of women, reinforced by upbringing, teachings of church and school, continuing inequality under the law, underrepresentation in government, and media portrayal of women as consumer queens. However, we must not underestimate the fact that there has always been a women's culture inside the patriarchal culture that modifies and changes patriarchal institutions. We can therefore think of women as shapers, not only victims, of society. However, individual initiative alone can be weak and ineffective. The support system for women found in women's organizations and networks acts as a great multiplier of individual effort. Below are three case studies that illustrate many of the ideas discussed in the preceding paragraphs.

■ SOME COUNTRY EXAMPLES OF THE WOMEN-DEVELOPMENT RELATIONSHIP

■ India

In 1974, in a remote village in northern India called Reni, nestled in the Garwhal mountains of the Himalayan range, a group of women and children were doing an unusual thing: hugging trees. They were resorting to *chipko,* which means to hug, to prevent trees from being felled. With their arms wrapped around the trees, the women and children cried: "The forest is our mother's home, we will defend it with all our might." The women and children stopped about seventy lumberjacks hired by forest contractors to fell the trees (Anand 1983).

With country after country selling off its forests to multinationals, not only displacing forest-dwelling people and removing their source of sustenance but destroying the very lungs of the earth and endangering the entire ecosphere, the Chipko movement has spread from village to village and country to country. It is mostly women's work to gather the rich supplies of forest plants, roots, berries, and tree fruits to feed their families and sell surpluses in the market for modest cash returns. It is their careful pruning of the trees for firewood that provides the only fuel for cooking. Therefore, women know they must be the protectors of the forests. They also know that trees help store water in the soil, that both water and the quality of soil itself will be lost when land is stripped of tree cover. And they know that financial resources needed for local development are in danger of being siphoned off to create roads and other infrastructure for an export-dominated economy, leaving the village poorer than ever. Already, women in the Gharwal area had to walk five to ten miles a day to collect firewood and fodder for cattle, bent double under heavy baskets.

In the local village councils, the leaders and decisionmakers are men, and these initiatives by women have startled them. Some villages are proud of their women and support them. Others know they are doing what is needed but feel that it is men who should be leading the way (except they do not). While there have been some sharp disagreements between women and men at the village level, the women have been able to sustain the courage of their convictions, supported by educated urban women like Vandana Shiva, director of the Research Foundation for Science, Technology, and Natural Resource Policy in Dehradun, India. Shiva has been an articulate voice linking care for the environment to a model of alternative development. She insists on the importance of building on grassroots capabilities that will remove the gulf between rich and poor. The village women now know they have a voice, and that voice is beginning to be listened to locally, nationally, and internationally.

■ Kenya

The Green Belt movement for reforestation is responding to a different phase of the same environmental crisis that the Chipko movement is responding to. The trees of the Rift Valley in Kenya are already gone, and the land is undergoing soil erosion and desertification on a large scale. The Green Belt movement brings the women of local villages into tree-planting activities in order to rebuild seriously degraded soils and also to provide the foods and fuelwood that a well-cared-for forest can provide. The plan is very simple. The women of each village plant a community woodland of at least 1,000 trees, a green belt. They prepare the ground, dig holes, provide manure, and then help take care of the young trees.

The inspiration for this movement came from a Kenyan woman biologist, Professor Wangari Maathai, who saw that the natural ecosystems of her country were being destroyed. As she has pointed out, the resulting poverty was something men could run away from, into the cities, but the women and children remain behind in the rural areas, hard put to provide food and water for hungry families. Working with the National Council of Women of Kenya, Professor Maathai went from village to village to persuade women to join the Green Belt movement. Green Belt by now has involved more than 80,000 women and a half million schoolchildren in establishing over a thousand local tree nurseries and planting more than 10 million trees in community woodlands, with a tree survival rate of 70 to 80 percent. This revival of local ecosystems has made fuel available, and family kitchen gardens are once more viable. Furthermore, women have been empowered to develop more ways of processing and storing foodstuffs that were formerly vulnerable to spoilage.

This movement, like the Chipko movement, has spread to other countries. In Kenya, as in India, educated urban women were in close enough touch with their village sisters to be aware of their needs. The basic development issue at stake, says Maathai, is food security, without which no country can have any meaningful development. Women helped to introduce the food security concept to development professionals.

■ *Bangladesh*

Long before the days of colonialism, the traditional practice of forming local credit associations had tided many women over hard times. The new situation of a growing cash economy in both rural and urban areas greatly increased the need for credit, but credit is rarely available to women through local banks. In Bangladesh, the Grameen Bank was started specifically for the poorest sector of the population and now has thousands of borrowers' groups throughout the country, three-fourths of them women. A network of roughly 8,000 "bankers on bicycles," each trained by the Grameen organization, covers many areas of Bangladesh, and the Grameen principle has now spread to other Asian countries, to Africa, and to Latin America. Further loans to a borrowers' group depend on repayment of previous loans, and repayment rates average 98 percent of all loans, with women having consistently higher repayment rates than men. Once a women's group has successfully completed its first round of loans for land for farming, for livestock, or for tools, they often expand their activities to building schools, clinics, and needed local production facilities. Studies of Grameen find that the incomes of borrowers' groups in Grameen Bank villages are 43 percent higher than that of borrowers in non-Grameen villages. This is a tribute both to the Grameen method and to the business acumen of women borrower groups.

■ ASSESSING WHERE WE ARE

We have followed the development decades from the perspective of women and noted the absence of women from the planning process. Male planners' lack of knowledge of the importance of women's agricultural labor, and of their economic and social contributions to a national standard of living, led development professionals and the UN development institutions, including the World Bank, to concentrate on a capital-intensive type of development. This included agricultural development that remained in special enclaves. Nothing "trickled down" to poor farmers, mostly women, and the countries of the South grew poorer as resources were diverted from the areas of greatest need.

We have also described how the international women's movement, with support from the UN, has set about to address the ignorance of male planners about women's work. At the same time, women's efforts to promote equality of participation in development, starting with the Women's Decade, has led to a growing awareness that overcoming poverty and improving the quality of life for all requires a different relationship between women and men, moving from patriarchy to partnership. When women and men can share their experience, resources of the World Bank and member states will be better used at local levels.

This is already happening. The World Bank has started modifying its structural adjustment programs. In the spring of 1997, a noteworthy global campaign was launched by INGOs and private sector antipoverty groups, with the backing of the World Bank, the UN Development Programme, national leaders, international aid agencies, foundations, and corporations, to make small loans to nearly 100 million poor people to finance small-scale farming and trade, with a special focus on working with local women farmers and entrepreneurs. Women's World Banking has been one of the leading promoters of this campaign, which will result in a tremendous multiplication of Grameen Bank–type projects. This represents an important step toward understanding development as human and social development, opening up more possibilities for diverse approaches to human betterment—recognizing, as Schumacher said a long time ago, that "small is beautiful." Most of all, it points to a growing partnership between the women and men who will be working for that more diversified, more earth-loving, more local and yet more connected world of the future.

■ DISCUSSION QUESTIONS

1. After reading this chapter, what does development mean to you?
2. Why have the international women's conferences from 1975 to 1995 continued with the same three themes: equality, development, and peace? Do you agree that these themes belong together?
3. The authors focus primarily on problems of development in the South. Do you think the United States has development problems?
4. What does the phrase *from patriarchy to partnership* mean? What would it mean for U.S. society?

5. What skills have the women discussed in the three case studies demonstrated in resolving current and future economic and environmental problems? How have they implemented the resolutions to these problems?

■ SUGGESTED READINGS

Boserup, Ester (1970) *Women's Role in Economic Development*. New York: St. Martin's Press.

Boulding, Elise (1992) *The Underside of History: A View of Women Through Time*. Revised edition. Newbury Park, CA: Sage Publications. (See vol. 1, ch. 4, "From Gatherers to Planters"; vol. 2, ch. 5, "The Journey from the Underside: Women's Movements Enter Public Spaces.")

Caldecott, Leonia, and Stephanie Leland, eds. (1983) *Reclaim the Earth: Women Speak Out for Life on Earth*. London: Women's Press.

Fisher, Julie (1993) *The Road from Rio: Sustainable Development and the Nongovernmental Movement in the Third World*. Westport, CT: Praeger.

Masini, Eleanora, and Susan Stratigos, eds. (1991) *Women, Households and Change*. Tokyo: UN University Press.

Shiva, Vandana (1989) *Staying Alive: Women, Ecology and Development*. London: Zed Books.

Turpin, Jennifer, and Lois Ann Lorentzen (1996) *The Gendered New World Order: Militarism, Development and the Environment*. New York: Routledge.

United Nations (1995) *Fourth World Conference on Women*. A/Conf. 177/20, October 17. New York: United Nations.

UNSD (United Nations Statistical Division) (2001) *The World's Women 2000: Trends and Statistics*. New York: United Nations.

Waring, Marilyn (1988) *If Women Counted: A New Feminist Economics*. San Francisco: Harper & Row.

CHILDREN

George Kent

Worldwide many children live in wretched conditions, suffering from malnutrition and disease, laboring in abusive work situations, and suffering exploitation of the most grotesque forms. The gravest problems of children are found in the third world, but even in industrialized nations many children are severely disadvantaged. In the United States, for example, fully one-fifth of the nation's children live below the official poverty line. My purpose here is to show that the situation of children should be understood not merely as a series of unconnected localized and private problems, but as systemic problems of public policy requiring attention at the highest levels of national and international governance.

Increasing attention by policymakers to the problems of children have resulted in some real progress in improving the quality of children's lives. The advances are documented every year in reports of the United Nations Children's Fund (UNICEF): *The State of the World's Children* and *The Progress of Nations*. In 1993, for example, *The Progress of Nations* reported,

> In little more than one generation, average real incomes have more than doubled; child death rates have been more than halved; malnutrition rates have been reduced by about 30%; life expectancy has increased by about a third; the proportion of children enrolled in primary school has risen from less than half to more than three quarters; and the percentage of rural families with access to safe water has risen from less than 10% to more than 60%. (UNICEF 1993a: 4)

However, satisfaction with such successes must be tempered with appreciation of the great distance still to be traveled if all children are to live a life of decency. Perhaps the clearest lesson learned in recent years is that significant gains in children's well-being do not result from economic growth alone. They also require progressive social policy based on a sustained commitment to improvements in the well-being of the poor in general and children in particular.

The following four sections provide an overview of the situation of children with regard to *child labor, child prostitution, armed conflict,* and *malnutrition.* A section on *child mortality* then shows that such pressures on children result in massive mortality, making even armed conflict look relatively unimportant by comparison.

While many different kinds of programs have been developed over the years to address the concerns of children, most have been inadequate to the task. There is now new hope in the rapidly advancing recognition of *children's rights,* based on the acknowledgment that every single child has the right to live in dignity. The legal obligation for the fulfillment of children's rights falls primarily on national governments, but for large-scale global issues these obligations need to be clarified. The last section of this chapter focuses on these obligations.

■ CHILD LABOR

Children work all over the world, in rich as well as poor countries. They do chores for their families, and many go out to fields and factories to earn modest amounts of money. Children's work can be an important part of their education, and it can make an important contribution to their own and their families' sustenance. There can be no quarrel with that. The concern here, however, is with child *labor*. Child labor can be defined as children working in conditions that are excessively abusive and exploitative. It is not clear where exactly the boundary between acceptable children's work and unacceptable child labor should be located, but there are many situations in which there can be no doubt that the line has been crossed. Abdelwahab Bouhdiba, in a study on child labor, offered many illustrations:

> Thousands of girls between the ages of 12 and 15 work in the small industrial enterprises at Kao-hsiung in southern Taiwan. . . . Some children [in Colombia] are employed 280 metres underground in mines at the bottom of shafts and in tunnels excavated in the rock. . . . Most carpet-makers [in Morocco] employ children between the ages of 8 and 12, who often work as many as 72 hours a week. . . . [In Pakistan] slave traffickers buy children for 1,600 rupees from abductors. They cripple or blind the weakest, whom they sell to beggar masters. . . . One million Mexican children are employed as seasonal workers in the United States. (Bouhdiba 1982: 2, 3, 11, 20)

Young brick carriers in Madagascar

Many children are caught up in the bonded labor system, especially in South Asia and Latin America. In the succinct explanation provided by the International Labour Organization's *World Labour Report 1993,*

> The employer typically entraps a "bonded" labourer by offering an advance which she or he has to pay off from future earnings. But since the employer generally pays very low wages, may charge the worker for tools or accommodation, and will often levy fines for unsatisfactory work, the debt can never be repaid; indeed it commonly increases. Even the death of the original debtor offers no escape; the employer may insist that the debt be passed from parent to child, or grandchild. Cases have been found of people slaving to pay off debts eight generations old. (ILO 1993: 11)

In Pakistan, an estimated 20 million people work as bonded laborers, 7.5 million of them children. The carpet industry alone has perhaps 500,000 bonded child workers. Afghan refugees in Pakistan, and their children, are now included in Pakistan's pool of bonded laborers.

Children work in rich countries as well. In the United States, for example, in 1988 about 28 percent of all fifteen-year-olds were working. The United States General Accounting Office (USGAO 1991) found that of the employed fifteen-year-olds, about 18 percent worked in violation of federal child labor regulations governing maximum hours or minimum ages for employment in certain occupations. Many working teenagers are injured on the job. Enforcement of child labor laws has been weak in many states, apparently due to the greater concern with protecting the interests of employers.

Paradoxically, the acceptance of child labor tends to be higher where there are higher surpluses of adult labor. The addition of children to the labor force helps to bring down wage rates, which in turn makes it more necessary to have all family members employed. The widespread employment of children keeps them out of school and thus prevents the buildup of human capital that is required if poor nations are to develop.

The United Nations Secretary-General has reported good progress in reducing child labor but acknowledges that there is much more work to be done:

> While most international attention during the 1990s was focused on the formal and export sectors, only 5 per cent of child labour is found there, and an estimated 70 per cent of children in developing countries work far from public scrutiny in agriculture and the informal sectors. The invisibility of the bulk of child labour—including work in the informal sector or in the family, represents a serious challenge and is compounded by the clandestine nature of such practices as trafficking. (UNGA 2001: 97)

Thus, despite extensive efforts to control the practice, there are still millions of children who work, many under grossly exploitative conditions.

■ CHILD PROSTITUTION

Child prostitution refers to situations in which children engage in regularized sexual activity for material benefits for themselves or others. These are institutionalized arrangements—sustained, patterned social structures—in which children are used sexually for profit. Child prostitution is an extreme form of sexual abuse of children and an especially intense form of exploitative child labor. Most prostitution is exploitative, but for mature men and women there may be some element of volition, some consent. The assumption here is that young children do not have the capacity to give valid, informed consent on such matters.

In some places, such as India and Thailand, child prostitution was deeply ingrained as part of the culture well before foreign soldiers or tourists appeared in large numbers. There are many local customers. Some Japanese and other tourists may use the child prostitutes in the "tea houses" in the Yaowarat district of Bangkok, but traditionally most of their customers have been locals, especially local Chinese. Similarly, in the sex trade near the U.S. military bases in the Philippines before they closed down, more than half the customers were local people. There is big money associated with the foreign trade, but there are bigger numbers in the local trade.

Child prostitution is widespread. It has been estimated that about 5,000 boys and 3,000 girls below the age of eighteen are involved in prostitution in Paris. The Ministry of Social Services and Development in the Philippines has acknowledged that child prostitution rivals begging as the major occupation of the 50,000 to 75,000 street children who roam metropolitan Manila. The number of underage prostitutes in Bangkok numbers at least in the tens of thousands. In India the number is surely over 100,000. It has been estimated that there are about 600,000 child prostitutes in Brazil. The UN Secretary-General's report on the status of children says:

> Sexual abuse occurs in the home, in communities and across societies. It is compounded when abuse takes place in a commercial setting. The worst forms of exploitation include commercial prostitution and child slavery, quite often in the guise of household domestic work. The trafficking of children, as well as women, for sexual exploitation, has reached alarming levels. An estimated 30 million children are now victimized by traffickers, so far largely with impunity. (UNGA 2001: 22)

■ ARMED CONFLICT

Armed conflicts hurt children in many ways. Wars kill and maim children through their direct violence. Children are killed in attacks on civilian populations, as in Hiroshima and Nagasaki. In Nicaragua, many children were maimed or killed by mines. The wars in Afghanistan in the 1980s and in Bosnia in 1993 were especially lethal to children. Wars now kill more civilians than soldiers, and many of these civilians are children. Children have been counted among the casualties of warfare at a steadily increasing rate over the past century. Historically, conflicts involving set-piece battles in war zones away from major population centers killed very few children. However, wars are changing form, moving out of the classic theaters of combat and into residential areas where civilians are more exposed.

There is also a great deal of violence against children in repressive conditions short of active warfare. Thousands of street children have been killed with impunity by death squads in Latin American countries.

Children are frequently hurt in the aftermath of warfare by leftover mines. The International Committee of the Red Cross has estimated that "using current mine-clearing techniques, it would take 4,300 years to render only twenty per cent of Afghan territory safe" (ICRC 1993: 471).

Often children are pressed to participate in armed combat as child soldiers, harming them both physically and psychologically. Children can be
the agents as well as the victims of violence. Increasingly, older children
(ten to eighteen years old) are engaged not simply as innocent bystanders
but as active participants in warfare.

Dorothea Woods, associated with the Quaker United Nations Office in
Geneva, has dedicated herself to chronicling the plight of child soldiers in
a monthly survey of the world's press entitled *Children Bearing Military
Arms*. In the January and February 1997 editions, for example, she cites
these cases:

- *Afghanistan:* "Hundreds of thousands of youth . . . were being raised
  to hate and fight a 'holy war'. . . . Many of those children are now
  with the Taliban army."
- *Chechnya:* "Government security forces have often detained young
  males between the ages of 14 and 18 as potential combatants in
  order to prevent them from joining the rebel forces."
- *Liberia:* "Because of the socio-economic crisis a part of the youth
  population is inclined to join one of the factions. The possession of
  a Kalachnikov gives the means to live by pillage and racketeering
  if necessary. . . . Various estimates have put the total number of
  Liberian soldiers below the age of 15 at around 6,000."

Small Boys Unit at a checkpoint near Lunsar, Sierra Leone

Photo credit: Sebastian Bolesch/Das Fotoarchiv

- *Sierra Leone:* "After the outbreak of the civil war in 1991 some five thousand youngsters joined either the governmental army or the rebel Revolutionary United Front."
- *Burma/Myanmar:* "A Shan boy . . . had been a porter-slave to carry heavy things to the place of fighting. . . . He fell down and was kicked by a Burmese soldier . . . until his leg broke like a stick in three places."
- *Guatemala:* "Forcing the under 18's from the indigenous communities to enroll in the army practically severs and destroys the future of these communities."
- *Mozambique:* "For the 10,000 children who took part in the civil war, the war is not over; it has been replaced by a multitude of small wars in their heads."
- *Uganda:* "Some 3,000 children have been kidnapped in the northern part of Uganda in the last four years according to UNICEF. The guerrillas who took these children have enrolled the boys in their army and have forced the girls to 'marry' the soldiers."

■ MALNUTRITION

Wars sometimes harm children indirectly, through their interference with normal patterns of food supply and health care. Many children died of starvation during the wars under the Lon Nol and Pol Pot regimes in Kampuchea (Cambodia) in the 1970s. In 1980–1986 in Angola and Mozambique, about half a million more children under five died than would have died in the absence of warfare. In 1986 alone, 84,000 child deaths in Mozambique were attributed to the war and destabilization. The high mortality rates in Angola and Mozambique were due not only to South Africa's destabilization efforts but also to their civil wars. The famines in Ethiopia in the mid-1980s and again later in that decade would not have been so devastating had it not been for the civil wars involving Tigre, Eritrea, and other provinces of Ethiopia. Civil war has also helped to create and sustain famine in Sudan.

The interference with food supplies and health services is often an unintended by-product of warfare, but in many cases it has been very deliberate. In some cases, the disruption of the infrastructure can have deadly effects well beyond the conclusion of the war. One example is the trade sanctions that have been imposed on Iraq after the Gulf War of 1991. More deaths resulting from the Gulf War occurred *after* the war than during it. On the basis of careful surveys in Iraq, in July 2000 UNICEF estimated that if trends of the 1980s had not been interrupted by the war and the subsequent sanctions, there would have been half a million fewer deaths of children under five in the 1990s.

There are many different kinds of malnutrition. One of the most important, protein-energy malnutrition, is usually indicated in children by growth retardation. It is widely accepted that if a child's weight is more than two standard deviations below the normal reference weight for his or her age (about 80 percent of the reference weight), that child should be described as malnourished.

An enormous number of the world's children are malnourished. The number is going down but much too slowly:

> In 1990, 177 million children under five years of age in developing countries were malnourished, as indicated by low weight-for-age. Estimates suggest that 149 million children were malnourished in 2000. The prevalence of under-five malnutrition in developing countries as a whole decreased from 32 per cent to 27 per cent. (UNGA 2001: 37)

The progress has been uneven:

> The most remarkable progress has been in South America, which registered a decrease in child malnutrition rates from 8 to 3 per cent. Progress was more modest in Asia, where rates decreased from 36 to 29 per cent and the number of underweight children under five years of age fell by some 33 million. Even this relatively limited achievement probably had a significant positive impact on child survival and development. Still, more than two thirds of the world's malnourished children—some 108 million—now live in Asia. . . . In sub-Saharan Africa, the absolute number of malnourished children has increased despite progress achieved in a few countries. (UNGA 2001: 37)

Contrary to the common belief that the problem is most widespread in Africa, there are far more malnourished children in Asia than in Africa. More than half the developing world's underweight children are in South Asia (UNACC/SCN 1997).

Almost a third of all children under the age of five in developing countries are malnourished, and malnutrition contributes to half the deaths of young children in these countries (UNICEF 1996a).

■ CHILD MORTALITY

Nothing conveys the plight of children worldwide as clearly as their massive mortality rates. Estimates of the number of under-five deaths for selected years are shown in Table 11.1.

Children's deaths account for about one-third of all deaths worldwide. In northern Europe or the United States, children account for only 2 to 3 percent of all deaths. In many less developed countries, more than half the deaths are deaths of children, which means there are more deaths of young

Table 11.1   Annual Child Deaths, 1960–2000

Year	Number of Deaths (0–5 years old)
1960	18,900,000
1970	17,400,000
1980	14,700,000
1990	12,700,000
1991	12,821,000
1992	13,191,000
1993	13,272,000
1994	12,588,000
1995	12,465,000
1996	11,694,000
1997	11,574,000
1998	11,140,000
1999	10,630,000
2000	10,929,000

Source: United Nations Children's Fund, *The State of the World's Children* (New York: UNICEF/Oxford University Press, annual).

people than of old people. The median age at death in 1990 was five or lower in Angola, Burkina Faso, Ethiopia, Guinea, Malawi, Mali, Mozambique, Niger, Rwanda, Sierra Leone, Somalia, Tanzania, and Uganda. This means that in these thirteen countries, at least half the deaths were of children under five. In the United States, the median age at death in 1990 was seventy-six, and in the best cases—Japan, Norway, Sweden, and Switzerland—it was seventy-eight (Kent 1995) (see Figure 11.1).

The child mortality rate for any given region is the number of children who die before their fifth birthdays for every thousand born. As indicated in Figure 11.1, the rate at which children are dying each year has been declining. However, the numbers are still enormous.

The number of children who die each year can be made more meaningful by comparing it with mortality due to warfare. There were about 100 million fatalities in wars between the years 1700 and 1987. That yields a long-term average of about 350,000 fatalities per year. The yearly average between 1986 and 1991 has been estimated at about 427,800. These figures can be compared to the more than 12 million deaths of children under five years of age in each of these years (Kent 1995).

The most lethal war in all of human history was World War II, during which there were about 15 million battle deaths. If civilian deaths are added in, including genocide and other forms of mass murder, the number of deaths in and around World War II was around 51,358,000. Annualized for the six-year period, the rate comes to about 8.6 million deaths a year—when children's deaths (under five years of age) were running at well over

Figure 11.1 Under-Five Mortality Rate, 1990 and 2000

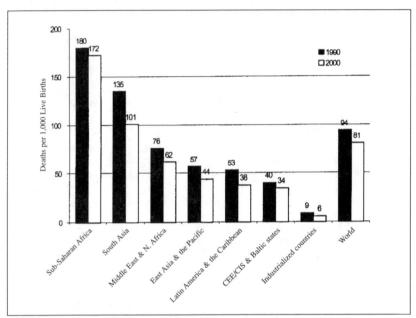

Source: Report of the UN Secretary-General, "We the Children: End-decade Review of the Follow-up to the World Summit for Children," May 4, 2001, A/S-27/1 (New York: United Nations).

20 million per year. This most intense war in history resulted in a lower death rate, over a very limited period, than results from children's mortality year in and year out.

Counting late additions, at the end of 1987 there were 58,156 names on the Vietnam Veterans' Memorial in Washington, D.C. That is less than the number of children under the age of five who die every two days throughout the world. A memorial for those children who die worldwide each year would be more than 250 times as long as the Vietnam Veterans' Memorial, and a new one would be needed every year.

Children die for many different reasons. The immediate cause of death for most children is not murder, direct physical abuse, or incurable diseases such as AIDS, but a combination of malnutrition and quite ordinary, manageable diseases such as diarrhea, malaria, and measles.

■ CHILDREN IN THE UNITED STATES

On the whole, children in the United States are far better off than most children in the rest of the world. However, there is a dark side that many of us never experience directly, one that is rarely reported to us by the government or the media. Although China may not be any more objective about this than Americans themselves, their views may provide a helpful corrective to our usual complacent views. The following excerpt from a report by the Chinese government is based entirely on data from U.S. sources.

Children in the United States live under worrying conditions, and they are often the major victims of violence and as many as 5,000 children are shot fatally annually. The percentage of gunshot victims under age 14 is 21 times that of 25 other industrialized countries. Some 1.5 million children, or two percent of the country's total, have one or both parents in prison. The United States, one of five countries that have the death penalty for juveniles, has the highest number of juveniles sentenced to death in the world. Twenty-five states of the country give death penalty to juveniles, four of which set the lowest age for the death penalty at 17 years and the other 21 states set 16 years as the bottom line or have no age limit at all. Since 1990, 14 juvenile criminals have been executed in the United States, and in the first seven months of 2000, four juvenile criminals were put to death, more than the figure of other countries combined in the past seven years. By October 2000, 83 juvenile criminals, who were under 18 when their crimes were committed, were waiting to be executed. The U.S. Department of Justice released a report on February 27, 2000, indicating that from 1985 to 1997, the inmates under age 18 in adult prisons more than doubled from 3,400 to 7,400; and 90 percent of juvenile criminals were high school dropouts. To date, more than 100,000 children are incarcerated in juvenile detention facilities and many of them are subject to brutal treatment.

Many children in the United States are threatened by poverty. According to an investigation conducted by the UNICEF, the poverty rate of children in the United States ranks second among the 29 members of the Organization for Economic Cooperation and Development. In 1998, the poverty rate of American children hit 18.7 percent, 2.5 percent higher than that of 1979. To date, as many as 13 million children live in poverty, three million more than the figure of 1979.

Reuters reported on January 20, 2000, that children in 15.2 percent of the families in the U.S. are starving, and that children aged below six years in 16.3 percent of households don't have enough food. About one million immigrant children who do not hold U.S. citizenship are not covered by the medical insurance system. More than one million children in the country live on the streets, 40 percent of whom are under 5, 20 percent suffer from hunger, 20 percent are not covered by the medical insurance system, 10 percent have seen murders, shootings, rapes and violence, and 25 percent have experienced domestic violence.

In the United States, at least 290,000 children are working in factories, mines and farms where working conditions are dangerous. Children working on farms often have to work 20 hours a day and run the risk of pesticide poisoning, injury and permanent disability. They account for 8 percent of the country's total child workers, while the job-related deaths among them make up 40 percent of the country's total occupational death toll. Among these child farm laborers, merely 55 percent have graduated from high school. It is estimated that there are one million cases of human rights violations against these child farm workers in the United States every year; yet the U.S. Labor Department listed only 104 such cases in 1998. (CIOSC 2001)

■ CHILDREN'S RIGHTS

Many different kinds of service programs are offered by both governmental and nongovernmental agencies to address children's concerns, and many of them have been very effective. However, the coverage is uneven, largely a matter of charity and chance. There is now an evolving understanding that if children everywhere are to be treated well, it must be recognized that they have specific rights to good treatment. Thus, there is now a vigorous movement to recognize and ensure the realization of children's rights.

Children's rights have been addressed in many different international instruments. On February 23, 1923, the General Council of the Union for Child Welfare adopted the Declaration of Geneva on the rights of the child. On September 26, 1924, it was adopted by the League of Nations as the Geneva Declaration on the Rights of the Child. It was then revised and became the basis of the Declaration of the Rights of the Child adopted without dissent by the UN General Assembly in 1959. The declaration enumerates ten principles regarding the rights of the child. As a nonbinding declaration, it does not provide any basis for implementation of those principles.

The Universal Declaration of Human Rights was approved unanimously by the UN General Assembly in 1948. It was given effect in the International Covenant on Civil and Political Rights and the International Covenant on Economic, Social, and Cultural Rights. The two covenants were adopted in 1966 and entered into force in 1976. The covenants include specific references to children's rights.

After ten years of hard negotiations in a working group of the UN Commission on Human Rights, on November 20, 1989, the UN General Assembly by consensus adopted the new Convention on the Rights of the Child. It came into force on September 2, 1990, when it was ratified by the twentieth nation. Weaving together the scattered threads of earlier international statements of the rights of children, the convention's articles cover civil, political, economic, social, and cultural rights. It includes not only basic survival requirements such as food, clean water, and health care but also rights of protection against abuse, neglect, and exploitation and the right to education and to participation in social, religious, political, and economic activities.

The convention is a comprehensive legal instrument, legally binding on all nations that accept it. The articles specify what states are obligated to do under different conditions. National governments that agree to be bound by the convention have the major responsibility for its implementation. To provide added international pressure for responsible implementation, Article 43 calls for the creation of a Committee on the Rights of the Child. It consists of ten experts whose main functions are to receive and transmit reports on the status of children's rights. Article 44 requires states parties to submit "reports on the measures they have adopted which give effect to the rights recognized herein and on the progress made on the enjoyment of those rights." Article 46 entitles UNICEF and other agencies to work with the committee within the scope of their mandates.

By the middle of 1997, all countries except Somalia and the United States had ratified or otherwise acceded to the Convention on the Rights of the Child. Somalia has not ratified because it does not have a functional government. The reasons for the U.S. failure to ratify are not so clear. Both Bill and Hillary Clinton were known as strong child advocates when Clinton first took office in 1993, so it was a serious disappointment to children's advocates when the Convention on the Rights of the Child was not quickly signed and ratified. The United States finally did sign the convention in February 1995. That signing, handled very quietly, apparently was done to fulfill a deathbed promise to James Grant, who had been executive director of UNICEF. However, the convention does not become binding on the United States until it is ratified through the advice and consent of the Senate. The convention still has not been forwarded to the Senate for its action.

The U.S. government has never offered any official explanation for its reluctance to ratify the Convention on the Rights of the Child. However, the major objections that have been voiced unofficially are described below.

States Rights

The historical struggle to find an appropriate balance between the powers of the states and the power of the national government has not been fully resolved. There is a fear that through its power to make international agreements, the U.S. government might federalize issues that previously had been addressed only in state law.

Capital Punishment

Article 37 of the convention states that "neither capital punishment nor life imprisonment without possibility of release shall be imposed for offenses committed by persons below eighteen years of age." Along with Pakistan, Iran, Iraq, and Bangladesh, the United States is one of the few remaining countries that execute people for crimes committed before their eighteenth birthday. This is tied in with the argument that capital punishment should be a matter of state rather than federal policy.

■ Abortion

The preamble of the convention says that "as indicated in the Declaration of the Rights of the Child adopted by the General Assembly of the United Nations on November 20, 1989, the child, by reason of his physical and mental immaturity, needs special safeguards and care, including appropriate legal protection before as well as after birth." The last six words conform to the pro-life, anti-abortion position. However, because of the divisiveness of the abortion issue, the drafters of the convention chose not to elaborate the theme. For pro-life activists, the convention is not explicit enough regarding safeguards before birth and thus is not acceptable.

It would be possible for the United States to ratify the Convention on the Rights of the Child despite these objections. It could be ratified with a reservation regarding the capital punishment provision, thus reserving U.S. freedom on that issue. There is nothing in the convention that would constrain the United States on the abortion question. Thus, there is no reason the United States should forgo supporting all of the other provisions of the convention that it favors because of these objections.

Perhaps the most serious obstacle to U.S. ratification is ideological. The United States tends to support civil and political rights but not economic and social rights. However, the Convention on the Rights of the Child asserts economic rights, such as the right to adequate nutrition. Article 24, for example, says that

> 1. States Parties recognize the right of the child to the enjoyment of the highest attainable standard of health and to facilities for the treatment of illness and rehabilitation of health. States Parties shall strive to ensure that no child is deprived of his or her right of access to such health care services.
> 2. States Parties shall pursue full implementation of this right and, in particular, shall take appropriate measures:
> (a) To diminish infant and child mortality;
> (b) To ensure the provision of necessary medical assistance and health care to all children with emphasis on the development of primary health care;
> (c) To combat disease and malnutrition, including within the framework of primary health care, through, inter alia, the application of readily available technology and through the provision of adequate nutritious foods and clean drinking-water, taking into consideration the dangers and risks of environmental pollution.

Economic rights of this sort trouble many people in the U.S. government. The government is willing to provide a broad array of nutrition and other social service programs, but it balks at the idea that people have a right, an entitlement, to these services.

Conservative elements in the United States have organized systematic campaigns of opposition to the convention based on false charges that the convention would undermine the family and take away parents' rights to raise their own children as they see fit. These unfounded arguments are advanced by people who apparently have not read the convention.

■ INTERNATIONAL OBLIGATIONS

Children have not only the rights enumerated in the Convention on the Rights of the Child but also, with few exceptions, all other human rights. The articulation of these rights in international instruments represents an important advance, but there is still much more to be done to ensure that these rights are fully realized. Although technically binding on the states that ratify these international human rights agreements, the human rights claims are not precisely specified. Latitude for interpretation is provided deliberately because it is left to the national governments, representing the states that are parties to the agreements, to concretize them in ways appropriate to their particular local circumstances.

Much more needs to be done by national governments to ensure that human rights within their jurisdictions are realized. However, the question remains: What are the obligations of the international community, especially where national governments are unwilling or unable to do what needs to be done to ensure that children's human rights are fully realized?

There are programs of international humanitarian assistance and many international organizations, governmental and nongovernmental, that work to alleviate suffering. Development and foreign aid programs do a good deal to improve the quality of life. But it is now largely a matter of politics and charity. There may be a sense of moral responsibility, but there is no sense of legal obligation, no sense that those who receive assistance are entitled to it and that those who provide it owe it. Historically, the idea of a duty to provide social services and to look after the weakest elements in society has been understood as something undertaken at the national and local levels, not as something that ought to be undertaken globally. Indeed, the only major market economy in which there is no clearly acknowledged responsibility of the strong with respect to the weak is the global economy.

Within nations, citizens may grumble when they are taxed to pay for food stamps for their poor, but they pay. Globally, there is nothing like a regular tax obligation through which the rich provide sustenance to the poor in other nations. The humanitarian instinct and sense of responsibility is extending worldwide, but there is still little clarity as to where duties lie. There is no firm sense of sustained obligation at the global level.

Most current discussions of global governance focus on security issues, the major preoccupation of the powerful, and give too little attention to the need to ensure the well-being of ordinary people. Just as there should be clear legal obligations to assist the weak in society at the local and national levels, those sorts of obligations should be recognized at the global level as well. Discussion of that idea has begun in the United Nations but just barely.

There is much discussion of international protection of human rights, but what does that mean? If one party has a right to something, some other party must have the duty to provide it. Children's rights would really be international only if, upon failure of a national government to do what was necessary to fulfill those rights, the international community was *obligated* to step in to do what needed to be done—with no excuses. There is now no mechanism and no commitment to do that. The international community provides humanitarian assistance in many different circumstances, but it is not required to do so. International law does not now require any nation to respond to requests for assistance.

There should be clear global obligations, codified in explicit law, to sustain and protect those who are the worst off. The exact nature of those obligations and their magnitude and form will have to be debated, but the debate must begin with the question of principle. The principle advocated here is that *international humanitarian assistance should be regularized through the systematic articulation of international rights and obligations regarding assistance*. Regularization can begin with the formulation of guidelines and basic principles and then perhaps of agreed codes of conduct. These can be viewed as possible precursors of law.

The nations of the world could collectively agree that certain kinds of international assistance programs *must* be provided, say, to children in nations in which children's mortality rates exceed a certain level. This international obligation to provide assistance should stand unconditionally where national governments or, more generally, those in power, consent to receiving the assistance. The obligation must be mitigated, however, where those in power refuse the assistance or where delivering the assistance would require facing extraordinary risks.

Part of the effort could focus on helping nations ensure that their children's nutrition rights are realized. The most prominent international governmental organizations (IGOs) concerned with nutrition are the Food and Agriculture Organization of the United Nations (FAO), the World Food Programme (WFP), the International Fund for Agricultural Development (IFAD), the World Health Organization (WHO), and the United Nations Children's Fund. They are governed by boards composed of member states. Responsibility for coordinating nutrition activities among these and other IGOs in the UN system rests with the Administrative Committee on Coordination/Sub-Committee on Nutrition (ACC/SCN). Representatives of bilateral donor agencies such as the Swedish International Development Agency (SIDA) and the United States Agency for International Development (USAID) also participate in ACC/SCN activities. There are also numerous international nongovernmental organizations (INGOs) concerned with nutrition. Some INGOs particpate in ACC/SCN activities.

The main role of the IGOs is not to feed people directly but to help nations use their own resources more effectively. A new regime of international nutrition rights would not involve massive international transfers of food. Its main function would be to press and help national governments address the problem of malnutrition among their own people, using the food, care, and health resources within their own nations. There may always be a need for a global emergency food facility to help in emergency situations that are beyond the capacity of individual nations, but a different kind of design is needed for dealing with chronic malnutrition. Moreover, as chronic malnutrition is addressed more effectively, nations would increase their capacity for dealing with emergency situations on their own. Over time the need for emergency assistance from the outside would decline.

The IGOs could be especially generous in providing assistance to those nations that create effective national laws and national agencies devoted to implementing nutrition rights. Poor nations that are relieved of some of the burden of providing material resources would be more willing to create programs for recognizing nutrition rights. Such pledges by international agencies could be viewed as a precursor to recognition of a genuine international duty to recognize and effectively implement rights to adequate nutrition.

Of course, the objective of ending children's malnutrition in the world by establishing a regime of hard international nutrition rights is idealistic. Nevertheless, the idea can be useful in setting the direction of action. We can think of the IGOs as having specific duties with regard to the fulfillment of nutrition rights. We can move progressively toward the ideal by inviting IGOs to establish clear rules and procedures they would follow *as if* they were firm duties.

■ CONCLUSION

Within nations, through democratic processes managed by the state, some moral responsibilities become legal obligations. A similar process is needed at the global level. Internationally recognized and implemented rights and obligations should not and, realistically, cannot be imposed. They should be established democratically, through agreement of the nations of the world. Reaching such agreement would be action not against sovereignty but against global anarchy. It is important to move toward a global rule of law.

Regularized assistance to the needy under the law is a mark of civilization *within* nations. If we are to civilize relations *among* nations, international humanitarian assistance also should be governed by the rule of law. Looking after our children internationally could become the leading edge of the project of civilizing the world order.

In 1990, a World Summit for Children produced ringing declarations and a promising plan of action to improve the conditions of children worldwide. In May 2002 a Special Session of the United Nations General Assembly was held to review the progress that had been made in the intervening decade and to make new plans and new commitments for the future. The Secretary-General's review, prepared for the Special Session, showed that substantial progress had been made on many of the issues of concern (UNGA 2001). It also showed that there was much more that remained to be done.

Like the 1990 summit, the Special Session of May 2002 concluded with impressive declarations and plans of action. Some will say that we can only wait and see if these commitments will be taken seriously. They are mistaken. It is important for people everywhere, in all walks of life, to actively and persistently insist that governments and international agencies at every level honor their commitments to the children of the world.

■ DISCUSSION QUESTIONS

1. Should countries be concerned with the treatment of children within other countries? Should the United Nations be concerned? Explain.
2. Should corporations be allowed to benefit from exploitative child labor in other countries? Explain.
3. Should the United States ratify the Convention on the Rights of the Child? Why or why not?
4. Do you agree with the concept of *international obligations* for children's rights?

■ SUGGESTED READINGS

CIOSC (China, Information Office of the State Council) *US Human Rights Record in 2000,* Beijing, February 27, 2001. Available online at http://www.chinadaily.com.cn/highlights/paper/us2000.html.
Freeman, Michael, and Philip Veerman, eds. (1992) *The Ideologies of Children's Rights.* Drodrecht, Netherlands: Martinus Nijhoff.
International Journal of Children's Rights (quarterly).
Kent, George (1991) *The Politics of Children's Survival.* New York: Praeger.
——— (1995) *Children in the International Political Economy.* London: Macmillan; New York: St. Martin's Press.

LeBlanc, Lawrence J. (1995) *The Convention on the Rights of the Child: United Nations Lawmaking on Human Rights*. Lincoln: University of Nebraska.

Sawyer, Roger (1988) *Children Enslaved*. London: Routledge.

United Nations (1992) *Child Mortality Since the 1960s: A Database for Developing Countries*. New York: United Nations.

United Nations Children's Fund (annual) *The Progress of Nations*. New York: UNICEF.

——— (annual) *The State of the World's Children*. New York: Oxford University Press/UNICEF.

——— (1993) *Food, Health and Care: The UNICEF Vision and Strategy for a World Free from Hunger and Malnutrition*. New York: UNICEF.

United Nations General Assembly (2001) *We the Children: End-Decade Review of the Follow-Up to the World Summit for Children: Report of the Secretary General*. New York: United Nations. A/S-27/3. Available online at http://www.unicef.org/specialsession/documentation/index.html.

Veerman, Philip E. (1992) *The Rights of the Child and the Changing Image of Childhood*. Dordrecht, Netherlands: Martinus Nijhoff.

HEALTH

Marjorie E. Nelson

The World Health Organization (WHO) defines health as a "state of complete physical, mental, and social well-being and not merely the absence of disease or infirmity." It is the result of a person's interactions with the environment. Genes, resources, past experiences, and choices determine some of the interactions humans have with their environment. The makeup of the natural environment is also a determinant. The complex interplay of all these factors results in a unique experience of health and disease throughout an individual's life. I examine some of these factors in this chapter. First, I look at nutritional status; second, I review examples of environmental factors; third, I examine some human behavioral factors; and finally, I consider some specific societal responses to protect health. Throughout the chapter I urge the reader to consider the disparity between health status in the North and the South.

■ NUTRITION AND HEALTH

"Phuong caught measles last week. He's much thinner than usual. His mother says he has no appetite and she thinks he's going to die," Madame Xuan Lan told me.

It was 1968. I was making my weekly visit to the day care center. Phuong, his mother, and three sisters were recent arrivals in one of the refugee camps around Quang Ngai City, where the American Friends Service Committee had set up this center during the Vietnam War. Phuong's mother supported her family by gathering wild greens to sell each day. She left Phuong in the care of his eight-year-old sister while she went to market.

215

While Xuan Lan went for Phuong, I examined other sick children. When Phuong arrived he was too weak to stand. I could hear pneumonia in his lungs and he already had a pressure ulcer on his lower back. His mother was right: lying without food or medicine on a damp floor in a refugee camp, Phuong would soon be dead.

We bandaged the ulcer and I left medicines for the pneumonia. "Bring him to school every day. Feed him chao thit ga three times a day and just let him rest." Chao thit ga is a Vietnamese chicken rice soup. It would give him the calories, protein, and fluids he needed to fight the infection, heal his ulcer, and regain weight. Sure enough, soon his pneumonia and ulcer were gone, and he was the bright-eyed four-year-old we had known before measles struck.

In North America, we do not think of measles as a killer. However, in developing countries, about one out of every four children who gets measles will die. The main reason is malnutrition.

■ INTERACTION OF PEOPLE, LAND, AND FOOD

Nutrition depends on what we eat. The interaction of people with the land at their disposal affects food availability and hence nutrition. When a people outgrows its food supply, one of three things will happen: Agricultural change will be introduced to produce more calories and protein per acre of arable land (land that can be cultivated); more land will be acquired—by migration or conquest; or people will starve.

In a traditional diet (a diet eaten by a people over a period of 500–1,000 years), the main source of calories is from grains and starchy foods: for example, wheat, potatoes, rice, beans, corn, plantains, and cassava. The highest-quality protein comes from animal sources: meat, milk, eggs, and seafood. However, other substantial protein sources for much of the world are soybeans, peas, beans, wheat, corn, rice, oats, rye, nuts, and especially peanuts. In traditional diets, fat is valued for its calories and taste but is a minor part of the diet.

To provide the yearly protein supply for a moderately active man by beef alone would require about 4.75 acres of land. To meet his yearly needs with poultry alone would require just under 2 acres. About 0.7 acre of wheat, 0.5 acre of rice, or 0.2 acre of soybeans would also meet his requirements. As the amount of arable land per person decreases, the usual diet in that region will shift toward food that has a higher protein content per acre.

Japan, a densely populated island nation, has a traditional diet that depends heavily on rice, soybeans, and seafood—which, coming from the sea, does not depend on arable land. Why? At the end of World War II, Japan had about 0.15 acre of arable land per person; this was not enough to meet Japanese protein needs, even if it were all in soybeans. Today it is about 0.12 acre per person. Japan, one of the richest and most developed countries, will always have to import food as long as its population remains as large as it is now. But the traditional Japanese diet tells us the country has had this problem for much longer than fifty years.

Rice was grown in Japan but was not significant in the Japanese diet until the seventeenth century. From 1600 to 1868, Japan underwent great economic development, and its population rose to 30 million. "Progress on this scale was possible only because of a constant rise in the agricultural production which supported these 30 million people on a surface area which could only have supported five or ten million people in Europe." Rice, a concentrated source of calories and protein, was the key (Braudel 1981: 156).

The typical U.S. diet is often described as "bread, meat, and potatoes." This could be a hamburger patty, bun, and fries; or fried chicken with mashed potatoes, gravy, and roll. At the end of World War II, the United States had 2.9 acres of arable land per person. Today it is about 2.1 acres per person. We are a nation of immigrants who have come over a period of 350 years to a place with abundant arable land. Our diet reflects that, as well as the diet patterns of our countries of origin.

■ UNDERNUTRITION

Adequate nutrition is crucial for a child's health. A good diet must include energy (calories), protein, and key micronutrients. Without adequate calories, a child's growth is stunted and weight is reduced. In case of infection, or in a time of food scarcity, a child like Phuong has no reserves on which to draw. The body needs protein for cell growth and repair. While calories are the fuel, proteins are the building blocks of the body: enzymes, hormones, cell walls, and antibodies to fight infections. Micronutrients such as iron, iodine, and vitamins are also important.

In 1995, according to the United Nations Children's Fund (UNICEF), 165 million children under age five were malnourished—half of them in just three countries: Bangladesh, India, and Pakistan (1996b: 12). Malnourished children, if they survive, grow into stunted adults with increased risk of disease, lowered capacity for physical and mental work, and, for women, reduced ability to nourish a developing fetus during pregnancy. Many of these effects cannot be totally reversed, even if nutrition improves later.

Now consider an example of migration and starvation. Food, as well as people, can migrate. The potato is a Western Hemisphere food. Grown in the high Andes since at least 2000 B.C.E., Spaniards brought it to Europe in the mid-1500s where it joined wheat (bread) as a major food. Soon it was widely grown in Ireland, and by the 1700s it was almost the entire diet of Irish peasants, augmented only by a little milk and cheese. By 1843, it was the single food sustaining the population of Ireland. Then disaster struck. A new potato blight appeared and wiped out the potato crop for two succes- sive years. For many reasons, adequate food was not brought into Ireland. Thousands died of starvation, while Irish immigration to North America in- creased greatly.

■ OVERNUTRITION

In the face of limited or sporadic food supply, *homo sapiens* developed the ability to store calories as fat when there is a temporary excess of food. Di- etary fat and carbohydrates can be converted to body fat and stored. Gram for gram, fat contains twice as many calories as carbohydrates, so it is an efficient energy storage system. In our distant past, persons with a prefer- ence for calorie-rich foods and an efficient system for storing extra calories as fat were more likely to survive, nourish offspring, and pass on both di- etary preferences and genetic traits. Today in some areas of the world, food surpluses, high incomes, and technology for the mass processing and dis- tribution of food make excess food available to many people. These foods are often high in fat and simple sugars and low in fiber compared to tradi- tional unprocessed foods such as fresh fruits, vegetables, and grains. As a result, obesity is a growing problem in most developed countries. Obesity is 120 percent or more of ideal body weight (see Table 12.1). Today nearly 33 percent of the U.S. population is obese.

Weight gain usually begins in childhood or early adulthood and con- tinues gradually throughout life. In recent years, lack of exercise has added to the risks from an unhealthy diet. Obesity and the diet that leads to it cause serious health problems: high cholesterol, high blood pressure, and hardening of the arteries. This results in heart attacks and strokes, the sec- ond and third leading causes of death in the United States. Gallstones and diabetes are more common in obese people. High-fat, high-calorie diets are also often low in dietary fiber. Dietary fiber, found in whole grains, fruits, and vegetables, slows the absorption of fat and sugar from the digestive system. This helps counteract a high-fat, high-calorie intake. Fiber affects bacterial metabolism in the digestive system, thus reducing the absorption of toxins. It also has a laxative effect. All these actions of fiber reduce the risk of some cancers, especially colon cancer. Cancer is the leading cause of death in the United States, and colon cancer is one of the top three along with breast and lung cancer. Once obesity occurs, like undernutrition, it is very hard to reverse the effects. The best strategy is to form good diet and exercise habits early and maintain them throughout life.

Table 12.1   Desirable Height and Weight for Adults

Height		Small Frame	Medium Frame	Large Frame
Men				
5 ft.	2 in.	128–134	131–141	138–150
5	3	130–136	133–143	140–153
5	4	132–138	135–145	142–156
5	5	134–140	137–148	144–160
5	6	136–142	139–151	146–164
5	7	138–145	142–154	149–168
5	8	140–148	145–157	152–172
5	9	142–151	148–160	155–176
5	10	144–154	151–163	158–180
5	11	146–157	154–166	161–184
6	0	149–160	157–170	164–188
6	1	152–164	160–174	168–192
6	2	155–168	164–178	172–197
6	3	158–172	167–182	176–202
6	4	162–176	171–187	181–207
Women				
4 ft.	10 in.	102–111	109–121	118–131
4	11	103–113	111–123	120–134
5	0	104–115	113–126	122–137
5	1	106–118	115–129	125–140
5	2	108–121	118–132	128–143
5	3	111–124	121–135	131–147
5	4	114–127	124–138	134–151
5	5	117–130	127–141	137–155
5	6	120–133	130–144	140–159
5	7	123–136	133–147	143–163
5	8	126–139	136–150	146–167
5	9	129–142	139–153	149–170
5	10	132–145	142–156	152–173
5	11	135–148	145–159	155–176
6	0	138–151	148–162	158–179

Source: Adapted from MetLife Insurance Company, www.metlife.com/lifeadvice/tools/ heightnweight/docs/men.html. Reprinted with permission of MetLife. This information is not intended to be a substitute for professional medical advice and should not be regarded as an endorsement or approval of any product or service.

Note: Weights at ages 25–29 based on lowest mortality. Weight in pounds according to frame (in indoor clothing weighing 5 lbs. for men and 3 lbs. for women; shoes with 1 in. heels).

■ HEALTH AND THE NATURAL ENVIRONMENT: NEW AND REEMERGING DISEASES

What we know as the AIDS pandemic began inconspicuously over forty years ago.

In 1959, two sailors died of a rare disease: pneumocystis pneumonia. One was in England; the other, a Haitian, was in New York. About that time, Dr. Margrethe Rask left Denmark to work as a surgeon in Zaire, and Gaetan, a boy in Quebec, dreamed of a career as an airline steward. In

1966, a Norwegian sailor died of multiple infections and a strange immune system collapse. Later, his wife and one child also died from results of severe immune deficiency. Two years later Robert, a teenager in St. Louis, died of multiple infections and Kaposi sarcoma. He had never been abroad, but he had been heterosexually active for "several years."

During the 1970s, Dr. Rask developed swollen lymph nodes and deep fatigue. Gaetan got a job with Air Canada and flew often between France, the United States, and Canada. He became sexually active as a homosexual. A young Greek fisherman moved to Zaire where he worked on Lake Tanganyika, while each year several thousand men from Haiti went to Zaire as short-term laborers before returning home. On July 4, 1976, New York harbor was filled with the tall sails of ships from fifty-five nations and their sailors who had gathered to help the United States celebrate its 200th birthday. It was a glorious party. In Zaire later that year, Dr. Rask collapsed with her strange illness and returned to Denmark. The Greek fisherman was not feeling as healthy as usual either. In late 1977, Dr. Rask died at the age of forty-seven with overwhelming pneumocystis pneumonia. In 1978, Dr. Peter Piot, a tropical disease specialist in Belgium, saw a new patient, the Greek fisherman from Zaire. The man died of widespread infection with an odd mycobacterium—one not recognized as a disease agent.

Meanwhile, Gaetan flew to San Francisco for the Gay Freedom Day parade. He enjoyed the visit and decided to return often. The next year, he experienced a "flu-like" illness that left him with swollen lymph nodes all over. He continued to fly between Paris, New York, and San Francisco, where he had many sexual contacts. In 1980, his doctor told him he had a rare cancer: Kaposi sarcoma. In Belgium, Dr. Piot continued to see patients from Zaire with severely damaged immune systems and strange infections: some women, some men, and some married couples.

On April 28, 1981, Sandra Ford, a technician at the Centers for Disease Control (CDC) in Atlanta, wrote her boss about a puzzling pattern. Part of her job was to fill orders for pentamidine, a drug used so rarely it was stocked only at the CDC. It was used to treat a rare pneumonia, pneumocystis, which typically occurred when a disease knocked out a patient's immune system, such as a child with leukemia or a patient with an organ transplant. Usually she got a dozen or so requests per year, but already this year she had nine requests, and the doctors did not seem to know why their patients had pneumocystis. Several were in New York City. Three weeks later, a doctor from Los Angeles called Dr. Mary Guinan at the CDC because he and his colleagues had a group of cases of pneumocystis pneumonia in gay men. They asked to have a report of this cluster appear in the CDC's weekly bulletin, MMWR. The June 5, 1981, issue of MMWR carried a report titled "Pneumocystis Pneumonia—Los Angeles." This was the first published report of a group of cases in the worldwide epidemic we have come to know as AIDS (based on Shilts 1987 and Garrett 1994).

Technically an epidemic is "the unusual occurrence of a disease in the light of past experience." Sandra Ford knew that even nine cases of unexplained pneumocystis pneumonia was a possible epidemic. When epidemiologists investigate a possible epidemic, they look for "Patient Zero." If they can find the first patient in the epidemic and learn about his or her interactions with other cases, they may find out how it spreads. If it is a new disease, it may help explain what causes it.

When CDC epidemiologists first began investigating what we now know as human immunodeficiency virus (HIV), the underlying cause of acquired immunodeficiency syndrome (AIDS), they did not know it was a virus. They also did not know how people became infected. They did know that it destroys part of the human immune system, allowing normally harmless microbes to invade and kill people. When they first found evidence that it was an infectious agent and was passed through sexual contact, they did not know how long a person could be infected before symptoms would occur. The first estimate was months. We now know a person may carry the virus and be infectious for more than ten to fifteen years before symptoms of AIDS appear.

In the early investigations, the CDC team found that Gaetan had been a sexual partner of several of the earliest cases in both New York and California. As they still thought the asymptomatic infectious period was months, he was tentatively labeled Patient Zero. He continued having unprotected sex even after being warned by several doctors that he was probably infectious. He certainly contributed to the spread of the AIDS epidemic, but he was not Patient Zero. Many of the earliest cases of AIDS detected in gay men in both New York and California were in a small group of men who lived and partied together in New York during the 1976 bicentennial celebration. Perhaps HIV was introduced to that group then, three years before Gaetan presumably got his infection.

Slowly we have put together the pieces of this AIDS puzzle. When HIV tests became available, doctors went back and tested samples of most of the early unidentified cases. They were HIV positive. We may never know where and how the AIDS epidemic started for sure, but clearly some cases occurred as early as 1959. However, the epidemic took off in the 1970s. Look at the cases before 1975: three sailors, the wife and child of one of them, a missionary doctor, a heterosexually active teenager who never left home, and a Greek fisherman working abroad. This is not the pattern we associate with AIDS today. But if someone had had all those cases collected then, that person could have warned us that an epidemic was coming.

Because air travel was uncommon before the 1970s, most of the early cases identified in the United States and Europe were travelers by sea; and their trips lasted days or weeks. Today we can fly from Miami to Managua in three hours, from Palo Alto to Beijing in thirteen, and from New York to Nairobi in fifteen. Almost half the world's people live in urban areas, and more than 10 percent live in cities of more than one million (WCED 1987a). What happens in one "neighborhood" of our global community can affect us rapidly, even though we live in another. In the developed world countries, AIDS began as a "big city" problem, apparently leapfrogging over small towns and rural areas to other big cities. Attention focused on gay men and intravenous (IV) drug users.

In Africa the pattern was different (see Table 12.2). In 1985, after over a year of trying to convince their colleagues and officials in Tanzania that they had AIDS in their small rural hospital on the Ugandan border, Drs. Kidenya, Tkimalenka, and Nyamuryekunge appealed to the CDC for help. They sent samples drawn from their patients to the CDC, which quickly confirmed the diagnosis of AIDS. Travelers were involved but here it was traders, truck drivers, or migrant laborers who had sex with prostitutes and then carried the disease back to their wives in rural areas (Garrett 1994). When I visited western Kenya in 1992, church elders told me of entire villages on the Ugandan border that were inhabited only by elderly grandparents and young children—orphans of parents who had died of AIDS. As epidemiological studies have emerged from other countries of the South, similar patterns have been demonstrated. AIDS is primarily transmitted heterosexually and from mother to unborn child, wiping out entire families and devastating communities. Sub-Saharan Africa is home to 28.1 million people infected with AIDS, about three-quarters of the total world cases. Africa has buried 75 percent of the world's AIDS victims (UNAIDS 2001).

As the world tries to cope with this new disease of AIDS, it also has to face the reemergence of diseases once thought under control. Humans are not the only ones interacting with the environment in our global neighborhoods; each neighborhood is home to many other living organisms: animals, plants, and microbes. Sometimes, as in the case of the HIV virus, what one microbe does improves the climate for other microbes. When HIV viruses have destroyed enough T-cells in a person's body, microbes, which are usually killed by our T-cells, can grow. Mycobacteria such as tuberculosis, parasites such as pneumocystis, or a new herpes virus that may be linked to Kaposi sarcoma are all examples.

Viruses and bacteria not only mutate—that is, change their genetic structure—but also exchange "tools" with other microbes. Many of these changes are useless or harmful to the microbe, but sometimes they help. One tool a bacteria might have is resistance to an antibiotic. Some bacteria carry the information on how to do this in "toolboxes" called plasmids, which they exchange with other bacteria. Before we discovered and started using antibiotics, this set of tools was not so useful, but in the past fifty years, bacteria that have one survive better.

Table 12.2   Regional HIV/AIDS Statistics and Features, 2001

	Adults & Children Living with HIV/AIDS (millions)	Adult Prevalence Rate[a] (%)	% of HIV-Positive Adults Who Are Women	Main Mode(s) of Transmission for Those Living with HIV/AIDS
Sub-Saharan Africa	28.1	8.40	55	Hetero[b]
North Africa & Middle East	.44	0.20	40	Hetero, IDU[b]
South & Southeast Asia	6.1	0.60	35	Hetero, IDU
East Asia & Pacific	1	0.10	20	IDU, Hetero, MSM[b]
Latin America	1.4	0.50	30	MSM, IDU, Hetero
Caribbean	.42	2.20	50	Hetero, MSM
Eastern Europe & Central Asia	1	0.50	20	IDU
Western Europe	.56	0.30	25	MSM, IDU
North America	.94	0.60	20	MSM, IDU, Hetero
Australia & New Zealand	.15	0.10	10	MSM
Total	40	1.20	48	

Source: UNAIDS, *AIDS Epidemic Update—December 2000*. Available online at http://www.unaids.org/epidemic_upsate/report_dec00/index_dec.html#full.

Notes: a. The proportion of adults (15–49 years of age) living with HIV/AIDS in 2001, using 2001 population numbers.

b. Hetero (heterosexual transmission); IDU (transmission through injecting drug use); MSM (sexual transmission among men who have sex with men).

Consider interaction of cholera bacteria with their environment. When we ingest water or food contaminated by cholera, they attach to the intestinal lining and reproduce. Unfortunately for us, they release a protein that causes severe diarrhea that can lead to death in days, or even hours.

Cholera has been present in South Asia for centuries. Periodically it would erupt as an epidemic across many countries, often carried by pilgrims traveling to Mecca. In the 1800s, there were many outbreaks in European cities due to inadequate sewage systems and unprotected water supplies. This virulent organism killed 70–80 percent of the people it infected, but it could not survive long outside the human body. Dr. John Snow, an early epidemiologist, stopped one outbreak in London by removing the handle from the pump at the Broad Street well—a cholera-contaminated water supply. Slowly, public health measures such as safe water, good sewage disposal, and use of quarantine reduced cholera epidemics.

Then, in 1961, a new strain of cholera, El Tor, appeared in Indonesia. It produced a milder disease; fewer ill people died, and some people were asymptomatic carriers. However, this strain was resistant to antibiotics. One substrain in Thailand was resistant to eight drugs. This cholera bacteria had picked up a powerful set of tools in a toolbox trade somewhere. It had also developed a way to survive longer in saltwater between stays in human hosts. These cholera organisms can hibernate for weeks at a time inside plankton that grow with algae. They can drift inside their hosts on the ocean currents or be drawn into a ship's bilgewater and pumped out in the next port of call a continent away. Both international shipping and increased algal blooms in the past three decades have helped the spread of cholera. A fiercer strain, Bengal, which appeared in the 1990s, combined some of the advantages of the El Tor with the more virulent disease-causing strength of the classic cholera.

During the past fifty years, we have lived through the Golden Age of Antibiotics. Now we are seeing new diseases like AIDS and reemergent old diseases with new defenses, like Bengal cholera, against which our antibiotics are no longer effective. In the next century, we must find new ways to deal with them and their toolboxes of tricks. This will require worldwide cooperation, as well as respect for changes that may arise in the environment as a result of our actions or other interactions.

■ HEALTH AND HUMAN BEHAVIOR

Above I pointed out that past experiences and personal choices determine some of the interactions humans have with their environment and that the social characteristics of that environment determine some of those interactions. Both affect health.

Rapid air transportation was not the only human behavior factor that helped the spread of the HIV/AIDS epidemic. The development of reliable birth control and of antibiotics that cured many of the recognized sexually transmitted diseases contributed to a sexual revolution among both heterosexuals and homosexuals. Most AIDS cases in the world today result from heterosexual sexual activity.

In contrast, the first cluster of AIDS cases identified in the United States was among gay men. It occurred during the tenure of a politically conservative administration in Washington that was not sympathetic to this group of citizens, nor to public health experts who tried to get more resources allocated to the emerging epidemic. The history of AIDS control efforts might have been very different if the first cluster had been detected in recipients of blood transfusions, for example.

Both individual actions and social structures or attitudes affect patterns of disease. These are not limited by national borders. Decisions in an Asian or Colombian neighborhood of our global community affect people in distant countries. During the Vietnam War, many soldiers were introduced to cheap heroin in Southeast Asia. The flow of heroin from there on both military and commercial flights rose dramatically. The 1970s saw the beginning of our current epidemic of IV drug use. In Latin America, a more powerful formulation of cocaine, "crack," was invented and began to spread among drug users. IV drug and crack cocaine users also have high rates of AIDS.

Not all drug traffic flows from developing to developed countries. Some is home grown and some flows the other way. In fact, the United States might be said to be the largest drug pusher country in the world if we consider the case of tobacco.

Lucy was a coworker and a patient of mine. We had known each other for years. Now she sat in my office with bad bronchitis.

"Lucy, have you ever thought about giving up the cigarettes?" I asked. She had smoked two packs a day since she was a teenager.

She swung one foot to and fro. "I've tried lots of times."

"Stop-smoking classes or nicotine gum have helped a lot of people quit," I suggested.

Her shoulders drooped. "I know, but I just can't quit. I guess I'll just have to die early."

I gave her a prescription for her bronchitis. I felt sad as I watched her walk down the hall. Her last words seemed ominous.

A couple of years later, Lucy had a heart attack and fell, breaking her hip. She died suddenly in the hospital from complications of the two events. She was fifty-two. Her words in my office that day had been prophetic: dying at fifty-two is quite early for a North American woman.

We are in the middle of an epidemic of disease and death caused by tobacco. It is a slow-growing epidemic compared to a cholera epidemic, which develops in days, or even to the AIDS epidemic. It takes thirty to thirty-five years for the damage caused by tobacco to show up as illness or death. Cigarette smoking causes lung cancer as well as heart attacks and thinning of the bones. Figure 12.1 shows that we really have two epidemics: one in men and one in women.

The epidemic began to show up in men in the 1940s but not in women until the late 1960s. Why? Smoking was not widespread in the United States before World War I (1914–1918). However, one of the standard items given to soldiers in World War I, along with equipment and food rations, was cigarettes. Every soldier got them. Many men came back with "the habit" and continued smoking. But it was a man's habit. It was not socially acceptable for women to smoke until World War II (1939–1945). So we see a twenty-five-year lag between the lung cancer epidemics in U.S. men and women.

Figure 12.1 Cancer Death Rates in U.S. Males and Females, 1930–1998

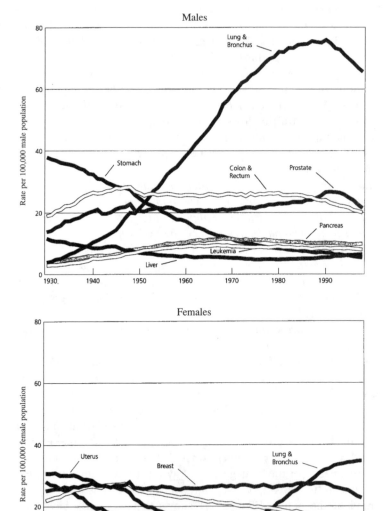

Source: American Cancer Society, *Surveillance Research, 2002.* Reprinted with permission of the American Cancer Society.
Note: Death rates per 100,000, age-adjusted to the 1970 U.S. standard population.

This same pattern is emerging around the world, due largely to U.S. and European tobacco companies. While tobacco is widely grown, in traditional societies it was the older men who used it. They grew it locally, cured it themselves, and "rolled their own" one at a time. If a man did not start smoking until he was a mature adult, he might live out his expected life span before he developed a cancer or died of other complications of smoking.

Today tobacco is the world's most widely grown nonfood crop and a significant item in international trade. According to the International Tobacco Growers Association, eighteen countries account for 80 percent of tobacco produced for export. The United States heads that list, exporting 30 percent of its total production. Asian countries, due to their rapid economic growth, are prime targets for U.S. cigarettes. Both the Reagan and George H. Bush administrations, for example, used the threat of trade sanctions to help open markets in Japan, Taiwan, and South Korea for U.S. tobacco companies.

In the United States, more money is spent on advertising tobacco products than on anything else except automobiles. In 1993, tobacco companies spent more than $6 billion on advertising and promotional items (Altman et al. 1996). The results? Nearly one in five U.S. high school seniors smokes daily, and about 40 percent of those smokers are female. Like Lucy, most adult smokers acquire their habit in their teens.

In China and Vietnam, where Western tobacco companies and their advertising are not yet well established, the traditional pattern predominates. Old men smoke and younger men are starting, but it is still uncommon among women, especially adolescent females. In Vietnam, only 4 percent of women smoke (Jenkins 1997), and in Beijing, only 3 percent of school-age girls smoke (Zhu et al. 1996). In countries where Western tobacco companies have introduced advertising and promotion campaigns, smoking patterns become more like those of the United States and Europe.

The health effects of this increase in tobacco use are enormous. In addition to lung cancer, tobacco use also causes many deaths from emphysema of the lungs. However, the greatest number of deaths due to tobacco use are cardiovascular: heart attack and stroke. Many of these deaths are premature—that is, before the age of sixty-five. Like Lucy, these people die in their forties and fifties. Tobacco use accounts for 27 percent of all premature deaths in the United States (Amler and Dull 1987). Already, lung cancer is the most common cancer in China. By the end of the decade, China will have 2 million smoking-related deaths each year; 900,000 of them will stem from lung cancer alone (Zhu et al. 1996).

Tobacco use is a good example of how both individual choice and social environment affect disease in our global community. International trade, advertising, social values in a "neighborhood," government policies, and the like, as well as an individual's decision, influence the magnitude and pattern of disease.

*   *   *

So far this chapter has looked at examples in three categories that influence health and disease in humans. The first category was nutrition. I described how Phuong's malnutrition left him with no reserves to fight against a serious illness. The second category was natural environment. I discussed how a new disease, AIDS, appeared and became a worldwide epidemic. New microbes, or reemerging old ones like Bengal cholera, with new tools to survive and spread better, present constant challenges to our health. The third category was social environment. I examined the cultivation, marketing, and use of tobacco products and how they have produced epidemics of lung cancer and increased the rate of heart attacks, strokes, and emphysema, which lead to additional illness and early death.

■ PROVIDING AND PROTECTING HEALTH: COMMUNITY ACTIONS

In this final section, I consider how our global village responds to health challenges. What do we as a community do to protect our health?

■ *Maternal and Child Health*

The nurses at Van Dinh District hospital were excited. Tuyet, a young woman from a small hamlet, had come to the hospital to have her first baby. Instead, she had given birth to three girls. Triplets are even more rare among Vietnamese than in the West. Everyone in the hospital wanted to see these three miracle babies. Although healthy, they were small so they would be in the hospital for several weeks. Dr. Thuat, the medical director, was pleased, not only because Tuyet and her daughters were healthy, but also because he saw an opportunity. When Tuyet went home, everyone would want to visit with her and the babies. She would be admired and respected because of these healthy triplets. Other young women would listen and take advice from her. So he went over to the obstetrics wing and invited Tuyet to enroll in their community health worker course. She could stay here with her babies while she studied. To his delight, Tuyet agreed.

Every community values its children, for they are the future. If children do not survive and grow into strong healthy adults, the community grows weak and risks extinction. So every community has developed activities to protect the health of children and their mothers.

Photo credit: Marjorie E. Nelson

These healthy eight-month-old triplets prove to be an attraction
that helps their mother, a community health worker, deliver
health education messages to her village in Van Dinh District, Vietnam.

As we have seen, nutrition is crucial to health. Adequate nutrition must begin before a baby is born. If a mother is poorly nourished, the developing fetus will not get all the nutrients it needs. The baby may be born under-weight, or even with birth defects. If a pregnant woman smokes, she may have a premature infant. If she drinks heavily, the baby may be retarded. Women need to know these things before they become pregnant, and they need access to good nutrition during pregnancy. In many countries, community health workers, like Tuyet, are trained to share this information with women in their neighborhood.

These health workers encourage women to breast-feed their infants. Breast milk is the best food for infants in their first four to six months of life for four reasons. First, it contains all the essential nutrients the growing baby needs except iron. Second, it contains antibodies to fight infections. A baby's immune system is not fully developed until about six months of age. Until then, it must depend on antibodies in mother's milk. The third and fourth reasons breast milk is best are that it is sterile and inexpensive. If a baby is fed with formula or other foods, there is a higher chance of illness, because poor families often stretch formula or other foods with contaminated water.

Sometimes community programs provide food supplements for preg-nant and nursing women and young children. In the 1980s, the government of Tamil Nadu in southern India began a program of weighing children every month from six months to three years of age. Any child who was malnourished received extra food for ninety days. At the same time, their mothers received nutrition education and were also offered extra food. When the program began, 45–50 percent of the children needed extra food. Eight years later that had dropped to 24 percent. "[This program] worked in Tamil Nadu because the community nutrition workers were well trained and highly motivated, and because mothers came to understand the importance of feeding for healthy growth" (World Bank 1993: 80). In the United States, this kind of community education work is often found in local health departments or community prenatal programs. At a county level, the Women, Infants, and Children (WIC) Program, funded by the U.S. Department of Agriculture, provides supplemental food to eligible women and children.

Some programs improve the nutrition and health of the whole population. Goiter and impaired thyroid function due to lack of iodine in the diet used to be quite common in the United States. It has disappeared, thanks to iodized salt. In the early 1900s, rickets caused by Vitamin D deficiency was very common in the northern United States and Europe. Now it is rarely seen because Vitamin D is added to milk.

■ Vaccines and Immunization Programs

Vaccination, or immunization, causes the body to produce antibodies against a specific bacteria or virus. This gives a person the ability to fight off an infection when exposed. In spite of his malnutrition, if Phuong had been immunized against measles, he would not have become sick and almost died.

In 1965, I heard a team of infectious disease specialists at the CDC present a plan to eradicate smallpox in ten years. The roomful of doctors from all over the world were skeptical. Never in history had humanity wiped out an infectious disease. Surely it could not be done in just ten years. In fact, it took them eleven.

On October 26, 1977, a Somali cook, Ali Maow Maalin, became the last case of wild smallpox in the world. This took a worldwide team effort. The United States contributed $30 million to the smallpox eradication campaign. Each year it saves about $360 million because it no longer has to vaccinate against, watch for, or treat any cases of smallpox in the United States. Barring a bioterrorist attack, smallpox will remain a distant memory.

The global community, through WHO and UNICEF, is now waging another ambitious campaign: the Expanded Program of Immunization (EPI) later renamed Global Alliance for Vaccines and Immunization (GAVI). When this program was begun in 1980, only 25 percent of the world's children were immunized against the six common vaccine-preventable diseases. By 1995, this had increased to 80 percent (UNICEF 1996b). One of these diseases, poliomyelitis, is slated for complete eradication by 2005. "The eradication of crippling poliomyelitis is 99% complete. . . . In 2000, there were no more than 3500 cases of polio reported worldwide" (UNICEF 2001). Perhaps polio will soon also be just a memory.

■ Influencing Human Behavior

There is no vaccine against AIDS. Combined treatment with two or more medicines prolongs the life of AIDS patients, but it is not yet clear that this can cure AIDS. These drugs may cost more than $3,000 per month per person. Most of the estimated 40 million people infected with HIV cannot afford that, so they will die early and without any substantial treatment. This is a great tragedy. However, an even greater tragedy will occur if we as a global community let the epidemic continue to spread. We now know how to prevent transmission of the HIV virus. The World Bank noted in its 1993 World Development Report that

> a comprehensive AIDS prevention program could check the growth of the disease. . . . Crucial elements in these strategies are providing information on how to avoid infection, promoting condom use, treating other sexually transmitted diseases, and reducing blood-borne transmission. . . . Young people, both in and out of school, need comprehensive education on reproduction and reproductive health issues. . . . All potential behavioral choices, including abstinence and condom use, should be presented. (World Bank 1993: 100)

In 1993, the global community spent about $1.5 billion on AIDS prevention. Schools, religious groups, voluntary agencies, and governments have all been involved in prevention programs.

No medical technology exists to undo the damage caused by tobacco use. By 2020, tobacco will kill more people than any single disease, even AIDS (Murray and Lopez 1996). However, it is another epidemic we know how to stop. Prevention can help both smokers and those who do not smoke. Smokers who stop cut their heart attack risk to the same as nonsmokers in one year and their added lung cancer risk in fifteen years. Lung destruction of emphysema stops when smoking stops. Of course, for those who never start, all these risks are avoided. Stop-smoking programs are offered by many hospitals, clinics, and health departments. Voluntary agencies, such as the American Heart Association and the American Lung Association, sponsor such programs as well as programs to prevent young people from starting to smoke or using smokeless tobacco.

Government policy also influences decisions to start or continue smoking. Actions that have helped reduce tobacco use include raising taxes on tobacco products, banning or limiting advertising, requiring warning labels on cigarette packages, prohibiting sales to minors, banning cigarette vending machines accessible to minors, and forbidding the sale of single cigarettes or distribution of free samples. Singapore has banned tobacco ads since 1971, and currently about eighteen countries have a total or partial ban on such ads. By the early 1990s, more than eighty countries required warning labels on cigarette packages. Tobacco taxes are fairly common. When Papua New Guinea raised the tobacco tax 10 percent, consumption dropped by 7 percent (World Bank 1993). These taxes can be used to support prevention programs.

■ Poverty and Health

So far I have examined some examples of specific activities designed to improve directly the health of people. Another factor mentioned earlier was the resources a person possesses. Poor people have more disease and poorer health than rich people. Studies show that this is true all over the world. With more money, people can buy more food, live in a place with safer water and sewage disposal, and get more education and medical care.

A poor country will usually have higher disease rates, higher death rates, and lower life expectancy than richer countries. Table 12.3 shows that, in general, as a country's per capita gross domestic product (GDP) falls, life expectancy falls, and death rates among infants and pregnant women and rates of tuberculosis rise. In a poor country, both individuals and the community at large lack resources to reduce death rates and improve the health status of people. "In 1984 few counties achieved an average life expectancy at birth of 70 years or more until gross national product per head approached $5000 per year" (Wilkinson 1992: 168).

Once a country reaches a certain level of economic development (about $5,000 per capita GNP in 1984), the *gap* between its richest and poorest people is a more important influence on the health of the nation than per capita GNP. As the gap between the richest and poorest gets smaller, life expectancy at birth increases. If the gap gets larger, life expectancy drops. This is based on studies in twelve countries of Western Europe and in the United States, Canada, Australia, and Japan. One might say, "as the rich get richer, the country gets sicker." Consider Great Britain and Japan:

> In 1970, income distribution and life expectancy were similar in these two countries. . . . Since then they have diverged. Japan has the most [equal] income distribution of any country on record [and] has the highest life

Table 12.3   Per Capita GDP and Selected Health Indicators, Selected Countries

Country	Per Capita GDP (PPP 1999 U.S.$)	Life Expectancy at Birth (1999)	Infant Mortality Rate (1999) (per 1,000 live births)	Maternal Mortality Rate[a] (per 100,000 live births)	Tuberculosis Cases (1998) (per 100,000 people)
Sub-Saharan Africa					
Nigeria	853	51.5	112	700	19
Kenya	1,022	51.3	76	590	169
Zambia	756	41	112	650	482
Niger	448	38.9	162	590	72
Middle East and					
North Africa					
Israel	18,440	70.6	6	5	10
Egypt	3,420	66.9	41	170	19
Yemen	806	60.1	86	350	73
Asia and Pacific					
Japan	24,898	80.8	4	8	35
China	3,617	70.2	33	55	36
India	2,248	62.9	70	410	115
Nepal	1,237	58.1	75	540	106
Americas					
United States	31,872	76.8	7	8	7
Brazil	7,037	67.5	34	160	51
Venezuela	5,495	72.7	20	60	27
Honduras	2,340	65.7	33	110	80
Europe					
Norway	28,433	78.4	4	6	5
Portugal	16,064	75.5	5	8	53
Russian					
Federation	7,473	66.1	18	50	82

Source: UN Development Programme, *Human Development Report, 2001* (New York: Oxford University Press, 2001).
   Notes: Purchasing power parity (PPP).
   a. For year available 1980–1999.

expectancy in the world. In Britain, on the other hand, income distribution has widened . . . and deaths among men and women aged 15–44 have increased. (Wilkinson 1992: 168)

The United States has the widest gap between poorest and richest families of any developed country. Government policies in terms of tax relief and other transfers can alter this gap. For example, without government intervention, France and the United States would both have about 25 percent of children living in poverty. After transfers and taxes, France's child poverty rate goes to 6.5 percent, while the U.S. rate falls to 21.5 percent. Just four other developed countries have child poverty rates above 10 percent, while ten have rates below 5 percent. Those ten all have lower death rates in infants and pregnant women, and all but Denmark have the same or higher life expectancy as the United States (UNICEF 1996b).

Health is one of the most precious possessions of any person. This chapter has considered some of the determinants of health: nutrition, poverty, the natural environment, the society in which we live, and the decisions we make individually or collectively. I have discussed examples of local programs that address these factors: feeding programs, immunizations, and stop-smoking classes. Both government agencies (such as health departments) and private voluntary agencies may deliver these services. Other factors can be operating in another neighborhood of our global community: emergence of a new virus, increased travel, tobacco trade, cocaine smuggling, government policies, and so on. Communications and cooperation between national and international agencies, such as the CDC, WHO, and UNICEF, play an increasingly important role in protecting our health when such factors are involved. No one nation could eradicate smallpox or monitor AIDS. Yet, a successful response to these challenges benefits us all.

■ DISCUSSION QUESTIONS AND SUGGESTED ACTIVITIES

1. How many examples of AIDS prevention programs can you identify in your own community?
2. If you saw the movie *Outbreak,* how does it demonstrate the influence of environment and human behavior on disease?
3. Should health care be a right of all people?
4. Should governments regulate the sale and purchase of tobacco products?
5. How do health problems differ in the North and South? Which problems are more serious? Which should receive the greatest attention?
6. Visit your local health department or voluntary agencies. Find out about their programs. How are they addressing any of the issues discussed in this chapter? Do they need volunteers?
7. Using Altavista or some other search engine, see what you can find on the World Wide Web about food and calories, the tobacco epidemic, or emerging infectious diseases.

■ SUGGESTED READINGS

Braudel, Fernand (1981) *The Structures of Everyday Life: Civilization and Capitalism—15th–18th Centuries.* New York: Harper & Row.
Crosby, Alfred (1972) *The Columbian Exchange: Biological and Cultural Consequences of 1492.* Westport, CT: Greenwood Press.

Garrett, Laurie (1994) *The Coming Plague*. New York: Farrar, Straus & Giroux.
Hobhouse, Henry (1987) *Seeds of Change: Five Plants That Transformed Mankind.* New York: Harper & Row.
McNeill, William H. (1987) *Plagues and Peoples,* New York: Anchor Books.
Shilts, Randy (1987) *And the Band Played On*. New York: St. Martin's Press.
Tannahill, Reay (1988) *Food in History*. New York: Crown Publishers.
UNAIDS (2001) *AIDS Epidemic Update—December 2000*. Available online at http://www.unaids.org/epidemic_update/report_dec00/index_dec.html#full.
UNICEF (2001) Press Release, "Polio Eradication: Final 1% Poses Greatest Challenge," April 3. Available online at http://www.unicef.org/newsline/01pr30.htm.
World Commission on Environment and Development (1987) *Our Common Future.* Oxford: Oxford University Press.

Part 4

THE ENVIRONMENT

13

PROTECTION OF THE ATMOSPHERE

Mark Seis

In this chapter I explore global warming, ozone depletion, and acid rain. I examine each separately, looking at the cause of the problem, the major nation-state contributors, and the various international polices that have been proposed and implemented for reducing the potentially damaging effects each poses for the future of humanity.

■ THE THREAT OF GLOBAL WARMING

Since Charles Keeling set up laboratories in 1958 at the South Pole and Hawaii, he has shown that carbon dioxide (CO_2) levels have been rising (Bates 1990; Leggett 1990; McKibben 1989). Current concentrations of CO_2 are still increasing. In fact, CO_2 levels are "according to samplings of air bubbles trapped in the world's deepest ice core in Vostok, Antarctica . . . 'unprecedented' in relation to the last 420,000 years" (Dunn 2001a: 85). There is widespread concern that increased CO_2 levels are leading to an increase in the earth's temperature, known as global warming.

Critics (Kerr 1989; Ray and Guzzo 1992; Michaels 1992; Lindzen 1993), however, question the actual amount of warming that will occur and the assumptions underlying computer model projections. Skeptics suggest that the earth's natural atmospheric process (for example, oceans, forests, sulfate aerosols) will be able to mitigate the greenhouse effect (described below). Other skeptics argue that the current warming trend is a natural fluctuation in global temperatures rather than a result of human activities.

Despite the critics, the majority of scientific literature on the subject is mounting a very strong case that warming (frequently referred to as climate change) is beginning to occur. The twentieth century was already 0.6 degrees Celsius (1.08 degrees Fahrenheit) warmer than the nineteenth century, and according to "global climate records . . . the 10 warmest years in the past century have all occurred since 1980" (Flavin 1996: 22). The most convincing evidence for global warming is the loss of ice sheets in the Northern Hemisphere. The thickness of arctic ice "has declined by 42 percent since the 1950s, and Norwegian researchers estimate that Arctic summers may be ice-free by 2050" (Dunn 2001a). The accelerated rate of warming has been attributed mostly to the activities of human beings (Bates 1990; Flavin 1996; Dunn 2001a) (see Table 13.1).

Table 13.1   World Carbon Emissions from Fossil Fuel Burning, 1950–2000

	1950	1960	1970	1980	1990	2000[a]
Emissions (million tons of carbon)	1,612.0	2,535.0	3,997.0	5,155.0	5,931.0	6,299.0
Carbon dioxide (parts per million)	n.a.	316.7	325.5	338.5	354.0	369.4

Source: Seth Dunn, "Atmosphere Trends," Worldwatch Institute: Vital Signs, 2001 (New York: W. W. Norton, 2001).
Note: a. Year 2000 figures are preliminary.

■ Causes and Consequences of Global Warming

The earth is constantly bombarded by solar radiation, some of which is absorbed and some of which is reflected back into space. This process is known as the greenhouse effect. When the natural carbon cycle is altered by fossil fuel burning (automobiles, power plants, industry, and heating being the most common fossil fuel burning activities), large amounts of CO_2 are released into the atmosphere. The result is more solar radiation is trapped in the earth's atmosphere and less is reflected, and therefore, the earth begins to warm. Gases like carbon dioxide, chlorofluorocarbons (CFCs), methane, tropospheric ozone, and nitrogen oxides trap solar radiation and cause the atmosphere to warm. Carbon dioxide accounts for roughly 50 percent of total greenhouse gases, chlorofluorocarbons 20 percent, methane 16 percent, tropospheric ozone 8 percent, and nitrogen oxides 6 percent (McKinney and Schoch 1996). As the atmosphere warms, it retains more water vapor due to evaporation. Water vapor is also a powerful greenhouse gas, because it traps long-waved solar radiation. This phenomenon has been recently demonstrated by satellite measurements from the Earth Radiation Budget Experiments (ERBE), which showed that as ocean and surface temperatures increase, more infrared radiation is trapped in the atmosphere (Leggett 1990).

Increasing temperatures set into motion various feedback loops that escalate the problem of global warming. A warming earth means that there is less ice and snow in mountain and polar regions to reflect back solar radiation. In addition, as the earth warms, large amounts of methane are released from ice, tundra, and mud in the continental shelves. More methane means more greenhouse gases to trap solar radiation, which in turn means hotter temperatures, and hotter temperatures mean more thawing, which means more methane gas—creating a vicious cycle of warming.

With global warming comes increased precipitation in some areas and drought in other areas because of increased evaporation due to the heat and changing wind patterns. Rising sea levels, longer and warmer summers, more severe storms, and more frequent forest fires are all likely outcomes of increased warming (Bates 1990; McKinney and Schoch 1996). Based on a plethora of global climate data, Flavin (1996) reports that sea levels are rising, summers are longer and warmer, serious storms have become more frequent and severe, and water shortages have become a chronic problem for eighty countries constituting 40 percent of the world's population.

Increased global warming will bring to extinction many more species before their natural time because of rapidly changing habitats (Leggett 1990; Weber 1993; Bright 2000). The resulting storm and drought damage may also lead to possible food shortages, especially when one considers the expected increase in human population.

■ Who Are the Major Contributors to Global Warming?

Since the Industrial Revolution, human activity has added 271 billion tons of carbon to the atmosphere (Dunn 2001a). Annual global carbon emissions are 6.3 billion tons (Dunn 2001a; 2001b). Since 1990, Western industrial nations have increased their carbon emissions by 9.2 percent, and developing nations have increased their carbon emissions by 22.8 percent (Dunn 2001b). Some countries are guilty of emitting much more carbon than others, which has created problems in the international community with respect to formulating binding international strategies for regulating greenhouse gas emissions. The largest producer of carbon emissions is the United States (1,433 million tons), followed by China (846 million tons), Russia (414 million tons), Japan (308 million tons), Germany (241 million tons), India (250 million tons), the United Kingdom (151 million tons), Canada (115 million tons), Italy (107 million tons), and South Korea (104 million tons) (Flavin and Dunn 1997). The United States also emits the most carbon per person (5.35 tons), followed by Australia (4.70 tons), Canada (3.82 tons), and Kazakhstan (3.29 tons) (Flavin and Dunn 1997). Three of the reasons stated for such high carbon emissions in the United States, Australia, and Canada are "low energy prices, large houses and heavy use of automobiles" (Flavin 1996: 31).

The fact that the United States has failed to take mandatory and legally binding steps to reduce its carbon emissions makes it difficult to persuade industrializing nations to slow their rate of carbon emissions and look to alternative technologies. Most U.S. measures to reduce the threat of global warming, as promulgated in President Bill Clinton's climate action plan, were voluntarily based. The Natural Resources Defense Council claims that even aggressive enforcement of Clinton's strategies would not have met target reductions because "in 1994 Congress approved only half the funds called for; the 1995 Congress made even more drastic cuts, and weakened appliance and lighting standards that had been enacted by the 1992 Congress" (Flavin 1996: 32). To date, President Bush has shown little concern in reducing carbon emissions in particular and global warming in general.

Hope is offered, however, by the fact that some industrialized nations such as Germany, France, the United Kingdom, and Russia have lowered their carbon emissions, and Denmark, Switzerland, and the Netherlands have implemented a variety of carbon emission reducing strategies (Flavin and Dunn 1997).

■ International Climate Control Policies

In 1972, the United Nations (UN) Conference on the Human Environment, or the Stockholm conference, was held. This conference consisted of 114 governments and was attended by many nongovernmental organizations (NGOs). It was the first time in history that nations of the world came together to discuss issues surrounding the destruction of the environment (Switzer 1994; Valente and Valente 1995). The Stockholm conference did not create any binding obligations but served more as a catalyst to generate an international discourse on global environmental issues.

It was not until the 1992 Earth Summit (also known as the Rio summit, United Nations Conference on the Environment and Development, or UNCED) that serious discussion on reducing CO_2 emissions to curtail global warming was undertaken. The Rio summit was attended by 178 countries and 110 heads of state (Switzer 1994). One of the five major documents produced at the Earth Summit was the Framework Convention on Climate Change. Its purpose was "stabilization of greenhouse gas concentrations in the atmosphere at a level that would prevent dangerous anthropogenic interferences in the climate system" (Flavin 1996: 36).

Although the United States, headed by the first Bush administration, was a major actor at the Rio summit, it fought binding targeted reductions in CO_2 emissions to 1990 levels by the year 2000. Despite U.S. reluctance to sign on to target CO_2 reductions, many industrialized European nations "did sign a separate declaration reaffirming their commitment to reducing their own CO_2 emissions to 1990 levels" (Gore 1992: xiv). Germany and

Japan provided most of the leadership in getting the other industrial nations to make a commitment to targeted CO_2 reductions. President Clinton in 1993 reversed the Bush administration's position, announcing that the United States would reduce CO_2 to 1990 levels by the year 2000. But as noted earlier, many of the appliance and lighting efficiency initiatives enacted by the 1992 Congress have been weakened by Congress in 1994 and 1995, which were also severely strained by receiving only half of the funding needed to implement many of the voluntary programs recommended by the Clinton plan (Flavin 1996).

Getting the industrialized nations to make serious commitments has become a major concern of the Alliance of Small Island States (AOSIS) and worldwide insurance companies. The AOSIS is a small coalition of island nations that are extremely threatened by rising seas. A rise of one meter in sea level could threaten to wipe out the sustainable land and economy of many of the small island nations. Given this threat, the AOSIS proposed that industrial nations reduce their CO_2 emissions by 20 percent. This proposal was also endorsed by seventy-seven other nations participating in the 1995 Berlin conference (which was a follow-up to the Rio summit and focused on climate change) but was resisted by a majority of the oil-producing nations like Kuwait and Saudi Arabia, and by the larger carbon-consuming countries like the United States and Australia (Brown 1996; Flavin 1996).

Other major nonnation players are some of the world's largest insurance companies, which obviously have a stake in any losses that may occur due to global warming. With sea levels projected to rise and temperatures projected to increase, insurance companies are reporting that "economic losses related to climate change could top $304 billion a year in the future" (Abramovitz 2001: 117).

The primary objective of the Berlin conference was to design measures that would reduce global carbon emissions and to create a series of trial projects aimed at exchanging alternative low-carbon–intensive technologies among nations (Flavin 1996). Despite the fact that there were no legally binding carbon reduction targets established, the Berlin conference did provide a sense of renewed hope in formulating a global policy for mitigating climate change. The agreement reached at Berlin, known as the Berlin Mandate,

> instructs governments to negotiate a treaty protocol "to elaborate policies and measures, as well as to set quantified limitations and reduction objectives within specified time-frames such as 2005, 2010, and 2020." (Flavin 1996: 35)

The meeting to establish this treaty took place at the end of 1997 in Japan and generated the Kyoto Protocol, which was adopted on December 11, 1997, and opened for signatures on March 16, 1998. The protocol contains legally binding emissions targets for key greenhouse gases, especially carbon dioxide, methane, and nitrous oxide. The agreement requires ratification by fifty-five countries that must include developed countries representing at least 55 percent of the total carbon dioxide emissions of those countries. According to the UN,

> The overall commitment adopted by developed countries in Kyoto was to reduce their emissions of greenhouse gases by some 5.2 percent below 1990 levels by a budget period of 2008 to 2012. While that percentage did not seem significant, it represented emissions levels that were about 29 percent below what they would have been in the absence of the Protocol. (UN 1998)

One of the problems with Kyoto was that negotiators were vague concerning the ways in which countries could achieve reductions through emissions trading and carbon sinks. In the former, countries would be assigned a total amount of carbon they are allowed to emit per year. These assigned levels vary for each country based on past emissions. If countries reduce their emissions below their allotted level, they can "trade" these unused emissions to countries having less success lowering carbon emissions. Kyoto was also vague on the degree to which carbon sinks should count toward a country's effort to reduce global warming. Carbon sinks include forests, rangelands, and croplands that absorb carbon. These details were to be worked out at a climate conference at The Hague, Netherlands, in November of 2000.

At the Hague conference the question of how carbon emission reductions could be counted remained the major obstacle to achieving an agreement. The United States, Canada, Japan, and Australia favored using flexible and creative approaches such as carbon sinks and market-based mechanisms (for example, emissions trading). Other European countries, however, interpreted the U.S. advocacy of flexible methods as an attempt to avoid reducing emissions from cars, factories, and power plants. Many European countries want actual reductions in greenhouse emissions through the development of non–fossil fuel technology. However, according to one study of five European nations only the United Kingdom and possibly Germany will meet the target emission reductions of Kyoto (Kerr 2000). It is unlikely that the United States will reach the emissions goals of Kyoto since according to one study it would have to reduce its emissions by 30 percent (Kerr 2000: 921). In contrast to Europe, the United States wants to achieve the Kyoto goals without reducing greenhouse gases, because such cuts would likely involve increased coal and oil prices. The Hague, Netherlands, negotiations ended in deadlock over agreement on how emission reductions could be measured.

In November 2001, negotiators from 160 countries met in Marrakesh, Morocco, and agreed on the details of the Kyoto Protocol begun four years ago. In the treaty, industrialized countries (with the glaring exception of the United States) agreed to reduce CO_2 by an average of 5.2 percent below 1990 levels. Countries have until the year 2012 to accomplish this task. Early in the year 2001, the industrialized countries decided to continue with negotiations even though the United States, under President George W. Bush, withdrew from negotiations. Even without the United States, negotiators believed they could get the required support of fifty-five countries, including those countries responsible for generating 55 percent of carbon dioxide emissions. The new agreement combines emissions trading and carbon sinks as well as required emissions cuts. Critics highlight that developing nations are exempt from this agreement (due to their economic situation and lack of contribution to the current CO_2 levels). They also question the significance of the treaty given the absence of the United States.

■ THE THREAT OF OZONE DEPLETION

Record low ozone levels were recorded throughout the 1990s. International policies appear to have reduced the sources of ozone depletion, but the damaging effects of a thinning ozone layer have yet to be fully realized, and it may be as long as a century before ozone levels return to concentrations prior to the creation of CFCs.

Causes and Consequences of Ozone Depletion

There are two types of ozone we often hear about. Ozone in the stratosphere is what protects us from harmful ultraviolet radiation. If compressed, the ozone layer would be about a tenth of an inch thick (3 mm). Ozone in the troposphere, on the other hand, is extremely poisonous to most life forms. A large portion of the ozone found in our troposphere is generated by human sources of atmospheric pollution that interact with solar radiation to create a pale blue gas with a strong pungent order, which can sometimes be smelled after it rains (Gribbin 1988). The issue highlighted in this chapter, in contrast, is the depletion of protective, stratospheric ozone resulting from the human production of chemicals like CFCs, halons, and other halon carbons. This stratospheric ozone layer protects us from excessive amounts of ultraviolet radiation.

The ozone layer is created from oxygen that escapes from the troposphere. Oxygen is created from living organisms that process carbon dioxide and exhale oxygen. Just as oxygen is vital for the creation of the ozone layer, so is the ozone layer vital for the creation of oxygen by protecting the living organisms that produce it.

CFCs and halon molecules, which have relatively long atmospheric lives, contain chlorine and bromine atoms. Once these molecules float their way up into the atmosphere, they interact with sunlight, which breaks apart their molecular structure releasing the chlorine and bromine atoms that destroy ozone. It is known that "a single chlorine atom can scavenge and destroy many thousands of ozone molecules" (Gribbin 1988: 48). John Gribbin explains:

> At an altitude of about 11 miles above the ground, more than half of the ozone above Antarctica was destroyed in the spring of 1987. And changes in the amount of chlorine oxide present marched precisely in step with changes in the amount of ozone. Where chlorine oxide went up, ozone went down, showing clearly that chlorine was destroying the ozone. (Gribbin 1988: xi)

CFCs are solely products of human industry and are most often used for propellants in aerosol spray cans. They are also used in air conditioners, refrigerators, computer chips, and styrofoam. Halons are used mostly in equipment used to suppress fires. There are no natural processes in the troposphere that react with CFCs and halons to break their molecular structure down (McKinney and Schoch 1996).

Another major destroyer of ozone is nitrous oxide (N_2O). Like CFCs and halons, N_2O is not a friend to the stratosphere. It reacts chemically with high-energy solar radiation to break up into NO (nitric oxide), which reacts with ozone. Nitrous oxide is emitted into the atmosphere by plants, combustion of coal and oil, and spray cans. Plants naturally emit N_2O, which helps maintain ozone levels so as to strike a balance between too little ozone—a certain amount is needed to protect the earth from harmful radiation—and too much ozone, which would inhibit life as we know it (Gribbin 1988; McKinney and Schoch 1996). While the release of N_2O by plants is a natural process, the use of chemical fertilizer to increase food production has increased the amount of N_2O in the troposphere (Gribbin 1988).

Effects of ozone depletion on human beings are several. The most obvious is that ozone depletion causes an increase in ultraviolet B (UV-B) radiation, which is known to increase our chances of getting skin cancer, especially malignant melanoma (Meadows, Meadows, and Rander 1992). Increases in UV-B radiation have been most pronounced in countries located in the Southern Hemisphere, such as Australia, New Zealand, and South Africa. Australia has the highest rate of skin cancer in the world, and researchers suggest that two out of three people growing up there will develop skin cancer during their lifetimes, "and 1 in 60 will develop the most deadly type, malignant melanoma" (Meadows, Meadows, and Rander 1992: 145). In the Southern Hemisphere, skin cancer and cataracts are increasingly common. In fact, Al Gore reports that

> in Queensland, in northeastern Australia, for example, more than 75 percent of all its citizens who have reached the age of sixty-five now have some form of skin cancer, and children are required by law to wear large hats and neck scarves to and from school to protect against ultraviolet radiation. (Gore 1992: 85)

In addition to increased rates of skin cancer, UV-B radiation also suppresses the human immune system, making us much more vulnerable to disease and viruses (Gore 1992; Gribbin 1988; Meadows, Meadows, and Rander 1992; McKinney and Schoch 1996).

Increased UV-B radiation on plants such as soybeans, beans, sugar beets, potatoes, lettuce, tomatoes, sorghum, peas, and wheat has been shown to inhibit growth, photosynthesis, and metabolism (McKinney and Schoch 1996). High levels of UV-B radiation also affect freshwater and marine ecosystems, especially ocean plankton, the base of the ocean's food chain (McKinney and Schoch 1996). There have been reports from Chile that sheep are going blind and that rabbits are developing myopia so severe that one can walk into a field and pick them up by the ears (Lamar 1991). Cattle are also known to get eye cancer and pinkeye when exposed to high levels of UV-B radiation (Gribbin 1988).

Major Producers and Users of Ozone-Depleting Substances

CFCs, like many other chemicals, really became popular after World War II. The production of the two most common types of CFCs increased from 55,000 tons in 1950 to 800,000 tons in 1976:

> From 1950 to 1975 world production of CFCs grew at 7% to 10% per year—doubling every 10 years or less. By the 1980s the world was manufacturing a million tonnes of CFCs annually. In the United States alone CFC coolants were at work in 100 million refrigerators, 30 million freezers, 45 million home air conditioners, 90 million car air conditioners, and hundreds of thousands of coolers in restaurants, supermarkets, and refrigerated trucks. (Meadows, Meadows, and Rander 1992: 142)

Approximately 90 percent of the CFCs used were immediately released into the atmosphere during their use, with the other 10 percent being released after the product was discarded (for example, refrigerators and air conditioners). Estimates are that 22 million tons of CFCs have been released into the atmosphere since their initial production (McKinney and Schoch 1996).

Like most pollution caused by modern technology, there is an immense disparity between the amount of ecological destruction caused by industrialized and industrializing countries. The countries of the North, such as the United States, the European Community (now the European Union), Japan, New Zealand, and Australia have used much more CFCs than industrializing countries like China and India (USCC and AN 1991). When broken down by per capita use, North Americans and Europeans were using on average 2 pounds (0.85 kg) of CFCs a year per person, and developing countries like China and India were using less than an ounce (0.03 kg) a year per person (Meadows, Meadows, and Rander 1992).

In 1985, there was no longer any doubt that CFCs were causing the destruction of the ozone layer. When British scientists measured a 40 percent decrease in ozone over Halley Bay in Antarctica, it was perceived as an error. After checking their measurement and other monitoring sites, it became apparent that ozone depletion was happening. NASA confirmed with readings made by the Nimbus 7 satellite that a hole in the ozone layer was indeed open over Antarctica. This discovery prompted a series of international meetings that led to a major conference designed to create target reductions and phaseouts of ozone-depleting substances.

■ International Policies on Ozone Depletion

Sherwood Roland and Mario Molina's 1974 paper documenting the depletion of ozone led to policies both in the European Community and in the United States requiring the phaseout of spray cans that used CFCs as propellants. In 1978, the United States banned the use of CFC propellants in spray cans, and the European Community reached a voluntary agreement requiring a 30 percent reduction in CFC propellant spray cans (Gribbin 1988; USSCEPW 1993). Despite the fact there were no international agreements at that time, there was a decrease in CFC propellants between 1974 and 1982 in most of the developed world except the Soviet Union (Gribbin 1988). The decrease in CFC propellants for spray cans, however, did not include a decrease in other uses and types of CFCs.

In 1985, the same year the ozone hole over Antarctica was discovered, an international agreement, titled the Vienna Convention for the Protection of the Ozone Layer, was signed by twenty nation producers of halocarbons. The agreement, however, did not entail any binding phaseout or reduction of ozone-depleting substances. Instead, the agreement focused on international research efforts aimed at documenting the ozone-depleting potential of halocarbons and other CFC and non-CFC ozone-depleting substances (Gribbin 1988; USSCEPW 1993).

After NASA's confirmation of the Antarctic ozone hole in 1987, it took only nine months of negotiations before twenty-seven countries signed the 1987 Montreal Protocol on Substances That Deplete the Ozone Layer. The Montreal Protocol is by far the strongest piece of international environmental policy to date. The agreement became effective January 1989 and required a "freeze in worldwide production of CFCs and halons (at 1986 levels, for CFCs in 1989 and halons in 1992) and a 50 percent reduction in the production and consumption of CFCs by mid-1988" (USSCEPW 1993: 108). The twenty-seven nations signing this document accounted for 99 percent of the producers and 90 percent of the consumers of ozone-depleting substances. Declining to sign, however, were many industrializing nations like India and China, because the original agreement made no stipulation for providing technical and financial assistance for developing nations (USSCEPW 1993).

Due to record-low ozone levels over the Northern Hemisphere reported by the scientific community in 1989 and again in 1992, the Montreal Protocol has been amended twice since its original signing. The first amendment took place in June 1990 in London. The amendment included an agreement by all the parties "to a complete phaseout of CFCs, halons and carbon tetrachloride by the year 2000" (USSCEPW 1993: 110). Further agreements included a ban in the year 2005 on the production of methyl chloroform. The London amendment also generated an agreement between the original Montreal Protocol signers to create a fund to provide assistance to developing nations for converting to the use of ozone-friendly substances (USSCEPW 1993).

A second major amendment was called for after the scientific community reported more record losses in ozone over the Northern Hemisphere (USSCEPW 1993). The second amendment was signed in November 1992 in Copenhagen by 126 countries. The amendment established more rapid phaseout dates for the major ozone-depleting substances. In addition to the phaseout of CFCs, hydrochlorofluorocarbons (HCFCs), the substance used to replace CFCs, are also targeted for phaseouts beginning in 2004. While HCFCs are much less devastating to the ozone layer, they are known to reach the stratosphere with 2 to 5 percent of the ozone-destroying potential of CFCs (McKinney and Schoch 1996). The full ramifications of HCFCs to the atmosphere and human health are not yet known, but the fact that the Copenhagen agreements call for a total ban on HCFCs by the year 2020 ensures that their overall impact on the environment will be mitigated.

It appears most countries are meeting their obligations because as of 1997 "global CFC production was down 85 percent from its 1986 level" (French and Mastny 2001: 181). Also, "a recent UNEP analysis found that at least two thirds of the developing countries surveyed were well on track to meeting their commitments" (French and Mastny 2001: 182).

Unfortunately, the only impediment to this treaty has been the illegal trafficking of CFCs. CFCs legally produced in developing countries have been making their way back into industrialized countries' black markets both in the United States and in Europe where CFCs are still sought after for their use in refrigerators and automobile air conditioners. Hilary French and Lisa Mastny note:

> By 1995, CFCs were considered the most valuable contraband entering Miami after cocaine. Following a subsequent crackdown on large consignments throughout East Coast ports, much of the illegal trade shifted to the Canadian and Mexican borders. Between April 1998 and March 1999, the U.S. Customs Office in Houston, Texas, reported 619 seizures of Freon, totaling nearly 20 tons. (2001: 182)

The good news, however, is the illegal trade of CFCs in the United States seems to be on the decline (French and Mastny 2001).

The Montreal Protocol and its amendments are testimony to the positive environmental policy that can be promulgated among nations when environmental degradation is taken seriously. It shows us that serious environmental degradation has a way of smoothing over ideological differences among nations. The Montreal Protocol and its amendments were passed because a diverse group of international scientists, politicians, and corporations agreed that preservation of the ozone layer—a necessity for life as we know it—outweighed the subtle political and economic differences that most often keep nations from international agreements. The Montreal Protocol and its amendments are the most successful international environmental policy to date.

■ THE ACID RAIN PROBLEM

Acid rain (often referred to as transboundary air pollution or simply air pollution) has become a major problem throughout the world, but it is most publicized in the United States, Canada, and Europe. Acid rain kills lakes and rivers, seriously damages the soil, and endangers the health of animals and humans. Like most pollution, acid rain does not stop at the political border of one country and ask permission to enter another. Acid rain has become a source of conflict between the United States and Canada and among many European nations.

■ Causes and Consequences of Acid Rain

Acid rain is created almost immediately after sulfur dioxide has been emitted into the atmosphere. Almost all fossil fuels contain sulfur, and sulfur content is extremely high in coal. In the United States, coal-fueled utility power plants account for roughly two-thirds of the U.S. sulfur dioxide emissions (Switzer 1994). When fossil fuels burn, sulfur combines with oxygen to create sulfur dioxide (SO_2), which is an odorless and colorless gas. Sulfur dioxide is a known lung irritant that can, in low concentrations, bring about asthmatic attacks and make those with respiratory problems quite uncomfortable. In the United States, sulfur dioxide by itself has not been a major problem, but the transformation of sulfur dioxide in the atmosphere into sulfuric acid (H_2SO_4) is a major environmental health problem (Seis 1996).

In the atmosphere, SO_2 combines with oxygen to form sulfate (SO_4). Sulfate is a small particle that floats in the air or settles on leaves, buildings, and the ground. When sulfate interacts with mist, fog, or rain it becomes acid rain (USCC and AN 1991). Sulfate, when inhaled into the moist lungs, is also transformed into sulfuric acid. All told, acid rain is dangerous to human health and to ecosystems in general, especially to trees, rivers, and lakes.

The disappearance of fish in the northern United States and in Canada has been on the increase since the early 1970s. According to one 1984 report, "In the Adirondack mountains, at least 180 former brook trout ponds will no longer support populations" (USCC and AN 1991: 3652). In 1975, another study, which surveyed 214 Adirondack lakes, showed that "90 percent of these lakes were entirely devoid of fish life" (USCC and AN 1991: 3652). The Office of Technology Assessment "estimated that in the Eastern United States approximately 3,000 lakes and 23,000 miles of streams have already become acidified or have virtually no acid neutralizing capacity left" (USCC and AN 1991: 3655). In Europe, "fish have disappeared from lakes in Sweden and Norway, as well as Scotland and England" (Switzer 1994: 258). Lake fish populations have been declining and in some cases dying out altogether in countries like Russia and Romania (Switzer 1994).

Acid rain has also been linked to forest declines in various parts of the United States. For example, in "Appalachia, the death rate of oaks appear to have doubled and that of hickories to have nearly tripled from 1960 to 1990" (Bright 2000: 34). Acidity in forest soil from Illinois to Ohio has led to "a decline of soil organisms—earth worms, beetles, and so on" (Bright 2000: 34). In Eastern Europe forests are dying rapidly, and in southwestern Poland, the army was used in 1990 to fell large tracts of dead forest due to acid rain. Military factories pumping out large doses of sulfur dioxide along the Russia-Finland border are responsible for ravaging forests within a 300-mile radius of the factories and are responsible for damage to "an additional fifty thousand square miles, with an estimated 30 percent of the firs in Finnish Lapland in danger of dying" (Switzer 1994).

The most serious acid rain problem in the world currently is in China. Seventy-three percent of China's energy comes from burning coal (Bright 2000: 34). Over a quarter of the country's land mass is now affected by sulfur dioxide (Bright 2000: 34).

The effects of acid rain on human health are beginning to make an appearance in populations throughout the world. In Santa Catarina, Brazil, for example, "the environmental secretary estimates that 80 percent of the local hospital patients have respiratory ailments caused by acidic pollutants" (Switzer 1994: 265). High acid levels have been correlated with increased colds, bronchial infections, asthma attacks, and death. Harvard public health researchers suggest that approximately a 5 percent annual excess of mortality in the United States is due to sulfate and fine particles (USCC and AN 1991).

■ Major Acid Rain Producers

As in the cases of ozone depletion and global warming, it is the largest industrialized nations that are the major generators of acid rain–causing pollutants. Coal-fueled power plants and factories account for the highest emissions of sulfur dioxide leading to acid rain worldwide. Thirty-nine percent of worldwide electricity is generated from coal (Flavin and Lenssen 1991). In the United States alone, 75 percent of total sulfur emissions come from power plants and large factories, which also emit large amounts of oxides of nitrogen (Switzer 1994).

The biggest problem with the countries producing the most sulfur dioxide and oxides of nitrogen emissions is that their pollution becomes the problem of other countries. Air pollution goes where the wind blows. Most of Canada's acid rain problems have been attributed to coal-fueled power plants operating in the Ohio Valley. Likewise, coal-fueled power plants and factories in eastern Germany, Poland, the Czech Republic, and the Slovak Republic destroy forests to the east. Sulfur emissions from Russia have been implicated in an acid rain problem for Finland and Sweden, and emissions from Great Britain have contributed to acid rain problems for Norway. Sulfur emissions from China create acid rain problems for Japan and South Korea.

The acid rain problem is severe in Europe; the International Institute for Applied Systems Analysis has estimated that "75 percent of Europe's forests are now experiencing damaging levels of sulfur deposition" (Brown 1993: 6). The monetary damage assessment for Europe is estimated to be at $30.4 billion per year, which does not take into consideration the other ecological functions forests serve with respect to regulating climate, flooding, and erosion (Brown 1993).

Given the fact that coal-powered electricity is the major culprit in acid rain generation, reducing our reliance on coal to generate electricity seems to be the obvious answer. No solutions between nations, however, are easy.

Coal is a cheap and widely available fossil fuel when compared with more expensive energy sources (for example, hydroelectric power, natural gas, and oil) and in some cases more dangerous alternatives (for example, nuclear power). Thus, solving what we know to be a simple problem becomes extremely complicated when we figure into the solution domestic and international politics and economics.

■ International Policies on Acid Rain

There have been no sponsored United Nations multilateral agreements regarding the abatement of acid rain. Unlike global warming and ozone depletion, which cause worldwide problems, acid rain is more regional. Accordingly, most acid rain agreements tend to be signed between bordering nations.

Probably the most extensive acid rain discussions have been between the United States and Canada. Canadian environmentalists contend that sulfur dioxide emissions from the United States are responsible for damaging as many as 16,000 Canadian lakes (Switzer 1994). Serious talks between Canada and the United States began in 1978 and culminated in a nonbinding treaty titled the Great Lakes Water Quality Agreement. The treaty stipulated measures that required emission reductions on the part of both nations. The progress made in 1978 was negated in 1980 when President Jimmy Carter, acting on the Middle East oil crisis, decided to convert 100 oil-fired utility plants to coal (Switzer 1994).

From 1980 to 1990, acid rain talks between the United States and Canada did not make much progress. It was not until 1990, when President George H. Bush signed the 1990 Clean Air Act (CAA), that serious efforts were initiated to reduce U.S. sulfur dioxide and nitrogen emissions.

With respect to sulfur dioxide emissions, the CAA promulgates a marketable allowance program. The allowance program grants each electrical utility a set level of emissions, which is roughly half of their emission output prior to the 1990 enactment of the CAA. Coal-fired utility companies can reduce their emissions below the EPA allowance and make money by selling their excess allowances to other utilities. Likewise, utilities can exceed their allowance by buying excess allowances from other utilities that have excess allowances to sell. Utility companies can profit by reducing sulfur dioxide emissions or pay for not reducing sulfur dioxide emissions (Seis 1996).

Europe has also been struggling with international strategies to curtail acid rain. The European Union (EU) and the UN Economic Commission for Europe (ECE) have been instrumental in at least initiating efforts to abate acid rain. The EU consists of fifteen member nations, and the ECE consists of all the nations of Europe. In an effort to make economic and environmental laws uniform, the EU has experienced difficulty in attempting to promulgate uniform sulfur dioxide emission reductions across nations. In fact, Great Britain, the largest producer of sulfur dioxide emissions in Western Europe, has often balked at emission standards despite being a country that has suffered from air pollution disasters (Switzer 1994).

The ECE has made more progress, beginning in 1979 with the enactment of the Convention on Long-Range Transboundary Air Pollution. This agreement of intent obliged each nation to develop technology to abate acid rain and where appropriate to share the technology. The agreement, however, established no emissions reduction standards nor did it require any uniformity in implementation of acid rain reduction technology. It was not until 1985 in Helsinki that thirty European nations agreed to a 30 percent reduction in sulfur dioxide emissions by the year 1993. Those that signed the protocol are known as members of the "30 percent club." While most European nations signed on, Great Britain, Spain, Ireland, Greece, and Portugal did not. Unfortunately, the 30 percent reduction is a politically derived target, not a scientifically based one, which means that there are no guarantees that a 30 percent reduction will abate the effects of acid deposition throughout Europe (Switzer 1994).

Countries like Norway, Sweden, Finland, and Japan have been making major inroads into generating sulfur dioxide–reducing technology and abating their own sulfur dioxide emissions. Due to the effects of acid rain on their own forests, Norway, Sweden, and Finland made Russia a $1 billion loan to utilize Finnish desulfurization technology (Switzer 1994). These same countries have been extremely influential in convincing most of Europe of the seriousness of acid rain. Japan has also been instrumental in helping the Chinese develop desulfurization technology. Unfortunately, many environmentalists contend that good desulfurization technology may not be enough for China, which has huge coal reserves and is just beginning to accelerate toward industrialized development (Brown 1993, 1996; Leggett 1990; Postel 1994; Bright 2000).

Many nations are recognizing the damaging effects of acid rain and are beginning to act by establishing sulfur dioxide emission reductions. The acid rain problem, however, is complicated for many reasons. Coal is the cheapest and most abundant fossil fuel remaining, and for developing nations it is at this time the only economically viable option for pursuing development. Even many of the more developed nations are reluctant to abandon coal as a major fuel. Although some reduction strategies are being implemented, it seems apparent that acid rain is not going to diminish as a major environmental problem anytime soon.

■ CONCLUSION

Of the first two problems discussed in this chapter, global warming appears on the international level to rank as less serious than ozone depletion. The major reason global warming appears to be a subordinate concern is probably due to the uncertainty that surrounds the possible outcomes of a warming planet. Furthermore, the causes of global warming are inextricably intertwined with politics, economics, growth, and development. Because global warming is taken less seriously, as judged by the commitment and legal teeth of international agreements, it could prove to be the world's most serious future problem. Radical climate unpredictability could create worldwide food and housing shortages. On a more optimistic note, however, current international efforts toward reducing greenhouse gases are much more serious than they were at the Rio summit.

International policy regarding ozone depletion is ecologically sound and moving toward a complete worldwide phaseout of major ozone-depleting substances. Unfortunately, it is going to be decades before international agreements produce major reductions in ozone depletion and a few decades more before ozone restoration becomes discernible. Nevertheless, the Montreal Protocol and its two amendments epitomize the type of international environmental agreements that can be achieved when nation-states recognize how they are interconnected ecologically.

Acid rain is also forcing nation-states to see beyond their political boundaries. Some countries are making major efforts to reduce emissions responsible for acid rain, but many nations are not because they are rich in low-grade coal and poor in desulfurization technology. Unfortunately, from an ecological standpoint, low-grade coal is the most available fossil fuel reserve for many countries, which means that acid rain will most likely continue to cause bioregional and geopolitical problems. The best hope for acid rain reduction worldwide lies in emission reductions of low-sulfur coal for industrialized nations and the easy availability of high-tech desulfurization equipment for all industrializing nations.

As this chapter has shown, atmospheric pollution influences the climate, and changing climate affects the fertility and integrity of the land. The next chapter examines ways in which atmospheric pollution is directly linked to land resources.

■ DISCUSSION QUESTIONS

1. What are some of the ways the largest greenhouse-contributor nations could reduce their emissions?
2. What are some of the major differences between the ozone problem and the global warming problem? Why is it so difficult for the world community to reach a viable solution regarding global warming, like they did with ozone depletion?
3. In what ways is acid rain an international problem? Describe some solutions to the acid rain problem if nations were to work together.

4. Does the North have an obligation to help the South develop in a more environmentally safe way? Does it have an interest in helping?
5. Should the South be expected to ratify the Kyoto Protocol?

■ SUGGESTED READINGS

Brown, Lester R., Christopher Flavin, and Hilary French, eds. *State of the World 2000*. New York: W. W. Norton.

———— *State of the World 2001*. New York: W. W. Norton.

Commoner, Barry (1990) *Making Peace with the Planet*. New York: Pantheon Books.

Dunn, Seth (2001) "Atmospheric Trends." In *The Worldwatch Institute Vital Signs 2001*. New York: W. W. Norton.

Leggett, Jeremy K. (2001) *The Carbon War: Global Warming and the End of the Oil Era*. London: Routledge.

Meadows D. H., D. L. Meadows, and J. Rander (1992) *Beyond the Limits: Confronting Global Collapse, Envisioning a Sustainable Future*. Mills, VT: Chelsea Green Publishers.

Schnaiberg, A., and K. A. Gould (1994) *Environment and Society: The Enduring Conflict*. New York: St. Martin's Press.

14

COOPERATION AND CONFLICT OVER NATURAL RESOURCES

Karrin Scapple

Wars have been fought and peace has been waged over natural resources. States rely on natural resources to sustain their economies and their independence. Since few states are self-sufficient, they often cooperate with other countries to obtain natural resources that they need; if cooperation is not possible, violent conflict becomes a viable alternative. The dilemma for a leader is to determine how to fulfill the state's needs with as little conflict as possible and without relinquishing too much state sovereignty. The resolution to this dilemma often depends on which natural resource is involved.

■ WHAT ARE NATURAL RESOURCES?

There are many types of natural resources and they have different impacts on global politics. Natural resources can be identified by whether they are renewable or nonrenewable and whether they stay within a single border or are transboundary.

■ Renewable Versus Nonrenewable Resources

A renewable resource is one that regenerates itself, such as trees, fish, and animals. Conversely, a nonrenewable resource is one that does not regenerate; once it is used, it cannot be recreated. In many respects *nonrenewable resource* is a misnomer. Most resources are, in fact, renewable. The issue becomes whether the resource can be renewed over a reasonable period of time in human terms. For instance, oil is a renewable resource and can regenerate over time; however, we would need to measure the regeneration

257

time in centuries, rather than months. So for policymaking purposes, it is more accurate to consider oil and other fossil fuels as nonrenewable resources.

Renewable and nonrenewable resources have different impacts on the international system. Theoretically, states should not have to fight over renewable resources because of their regenerative characteristics. If states are unable to meet their own needs for a particular resource, they are likely to cooperate to meet those needs through trade agreements and economic integration. Conflict results, however, if the needed renewable resource is overconsumed so that full regeneration, or sustainable growth, is no longer possible. Fishing and whaling conflicts have become key issues over the past several decades.

Conflict and violence sometimes result over nonrenewable resources. If a resource is needed but there is a finite amount available, states will sometimes fight very hard to obtain that resource. The Gulf War (1990–1991) is an example of a highly visible clash over nonrenewable resources. While cooperation is possible, it becomes less likely if the nonrenewable resource is, like oil, critically needed and if the disputants have unresolved conflicts from the past.

■ Boundary Versus Transboundary Resources

The question that must be addressed here is whether the resource stays in one place or whether it moves. A forest is an example of a boundary resource; most are located only within one state's borders and ownership is clear. Conflict is less likely with a boundary resource because of the international principle of sovereignty. A river, however, is an example of a transboundary resource that may either define a border between two countries or travel from one country to another; in either case, the river must be shared by two or more states. Although the opportunity for cooperation increases in a transboundary resource, so does the possibility for conflict.

The issue of sovereignty becomes critical when one looks at natural resources in boundary and transboundary terms. International law protects a state's sovereignty and its territorial integrity. International law asserts that the resources found within a state's borders are that state's property to do with as it chooses. Although countries have concerns over Nigeria's human rights abuses involved in oil development, no country disputes Nigeria's right to develop its oil fields. The oil fields are in Nigerian territory and are under Nigerian sovereignty.

Yet this sense of sovereignty becomes controversial with transboundary resources. If a river head exists in one state, does that state have the right to do whatever it wants with the water, even though downstream states may be dependent on that resource too? The answer to this question depends on whether a country is "upstream," or "downstream." Upstream states tend to rely heavily on the sovereignty principle; downstream states tend to promote the idea of equity and cooperation. Cooperation may be relatively easy over some resources, but conflict becomes more likely if the resource is critical to survival.

■ CASE STUDIES

I have created a fourfold matrix of issues using the factors of renewability and location (see Table 14.1). The remainder of this chapter will focus on a case study in each of the cells of this matrix in order to highlight some natural resource issues. These cases provide examples of how natural resources can be seeds of cooperation as well as sources of conflict in the international community.

Table 14.1   Natural Resource Matrix

	Renewable	Nonrenewable
Boundary	Forests	Oil
Transboundary	Fish	Water

■ Renewable, Boundary Resource: Forests

Again, the use of a boundary resource, such as a forest, sits clearly within the domain of the sovereign state because the resource does not travel outside its borders. While this is true, use of a boundary resource can often have second-order consequences for surrounding states or for the global community at large. For instance, even though a forest does not travel, a river that runs through a forest does; if that forest is clear-cut, there will be enormous ramifications on the river as a result of sediment increase or flooding. Further, that forest's contribution to regional and global climate conditions will be lost, with possible serious implications. As a result, the international community has come to be concerned about a state's use of its resources when there are possible global consequences.

There are debates about who should have jurisdiction over a boundary resource when there are second-order consequences. The North is currently advocating that the needs of the international community should be a priority. Issues such as global climate change, stratospheric ozone depletion, and law of the sea should take precedence in policymaking decisions. The South, however, is primarily concerned about its own development opportunities. Boundary resources may provide a good source of export income and meet internal development needs. These states believe they have the right to develop and meet the needs of their people. Further, the South believes that environmental concerns are being used as yet another way to ensure that the South does not develop and fully compete with the industrialized North.

The irony is that not only do many of the development patterns in the South actually limit future potential for growth but the North makes a great contribution to habitat destruction. The nutrients in most tropical forests are found in the vegetation, not the soil; a forest that is clear-cut can be productive for only three to five years. Thus, it may be in the South's best interests to find other ways of developing, rather than through deforestation. But the North has some culpability in this process. U.S. and European demands for wood products, as well as the penchant for inexpensive beef, have created the trade environment that encourages deforestation. The Amazon rain forest has been a particular concern because of its potential to help reduce atmospheric carbon dioxide. Yet, over the past thirty years, the trade in forest products has increased dramatically, with exports from South America going primarily to the United States. By 1998, the United States was the largest importer of Chilean wood products of both native and nonnative species. Today, the U.S. import of Brazilian wood products, primarily pine (a nonnative species) is six times higher than it was in 1990 (World Forestry n.d.). Similar trends exist with the cattle trade. Brazil and Argentina are the fourth and fifth largest exporters, respectively, of beef to the United States. In fact, Brazil's cattle industry is currently half again as large as its U.S. counterpart (Public Policy Center 2000).

While Brazil, Chile, and Argentina have the sovereign right to use their boundary resources as they desire, the second-order consequences for the global community may be dramatic. When the North, particularly the United States, suggests that the South should protect the environment, the North must also make environmentally friendly changes in its lifestyles.

The dilemma is to determine how forests can be used for development and at the same time continue to make their contributions to the ecosystem.

Debt-for-nature swaps in Costa Rica. One solution might be debt-for-nature swaps, which have met with some success in Costa Rica. More than 25 percent of Costa Rica, a country about the size of West Virginia, consists of parks and protected areas. Yet outside these protected areas, Costa Rica has one of the highest deforestation rates in the world, at almost 3–4 percent annually (Jones 1992; Sarkar and Ebbs 1992). It is estimated that only 20 percent of original, primal forest remain (Baker 1999). Deforestation has occurred because of the combined pressure of population growth and industrialization. Furthermore, due to its foreign debt, it has been estimated that Costa Rica might have to use 25 percent of its annual budget to service its international debt (Sarkar and Ebbs 1992). Clearly, supporting a forest would be considered a "luxury." The strain of development, combined with a decentralized government, caused many of the parks to be undermanaged by government agencies and penetrated by individuals in search of meeting basic human needs. Costa Rica has found that debt-for-nature swaps are one way to lessen the problem, and it has been the most active state in terms of the number of swaps, the amount of money generated, and the amount of debt reduced (Mahony 1992).

Cooperation is the key word in debt-for-nature swaps. Several actors are involved: financial institutions, nongovernmental organizations (NGOs), and states. These actors negotiate an arrangement that reduces the state's international debt while also preserving some of its forests for conservation and sustainable development. Most states that are heavily deforesting are also heavily in debt: Costa Rica is only one such state. While each case differs, the general arrangement is that an NGO in the North (Conservation International, World Wildlife Fund, and Nature Conservancy have been the most prominent) raises funds to buy-down a portion of the state's international debt at a reduced rate. This arrangement is made through negotiations with the state that owes the debt and the financial institution; in Costa Rica's situation, one buy-down was negotiated at only 17 cents on the dollar (Baker 1999). In exchange, the government agrees to provide a local NGO with the equivalent amount in local currency or through bonds. With this monetary commitment, the NGO is able to create a park or protected area out of what would have become deforested land; many of Costa Rica's protected areas have been created through these arrangements. It appears to be a win-win situation: The financial institution gets something for a debt that may not otherwise be repaid; the state gets part of its international debt eliminated; and the NGOs help save part of a natural forest that can continue to contribute to the local and global ecosystem.

Debt-for-nature swaps received new prominence in 2001 when the U.S. government agreed to participate, with the Nature Conservancy, in a swap expected to reduce Belize's debt to the United States by one-half. Further, approximately 23,000 acres of forested land, including 16 miles of pristine coastline, will be protected by the monies provided to four Belizean NGOs. This is the first time that a private organization has negotiated a debt-for-nature swap with the U.S. government under the recently passed Tropical Forest Conservation Act (Nature Conservancy 2001).

While there are many benefits of the debt-for-nature swaps, there are also disadvantages. Many banks are unwilling to sell the debt at a reduced rate. In fact, the World Bank and other multilateral development banks are prohibited from doing so; thus, about 60 percent of the global debt is beyond the reach of the debt-for-nature process (Klinger 1994). Even with the eligible debt, many critics are concerned that only a small amount of the debt is actually eliminated (Mahony 1992). Costa Rica, the model for the debt-for-nature concept, has been able to eliminate only 5 percent of its international debt. There are continuing concerns that the parks created are still not well managed and that deforestation persists despite the transaction (Sarkar and Ebbs 1992). Debt-for-nature swaps are much more difficult to negotiate when debtor states are politically unstable (UNDP 1998). Finally, some argue that it is still the North that benefits most from the arrangement and that it has domination over the process (Mahony 1992).

Despite the controversy, debt-for-nature swaps offer a unique way to resolve natural resource issues. They provide an opportunity for cooperation at many levels: between North and South, between bank and debtor, and between governments and nongovernmental organizations. With some modifications to the system, debt-for-nature swaps have become a valuable model for other actors, besides environmental NGOs, to help resolve conflict on a wide variety of other issues (Michaels n.d.).

■ Renewable, Transboundary Resource: Fish

Any child who has owned a couple of guppies can tell you that fish are a renewable resource. In fact, fish renew at a very fast rate. Yet, according to the UN Food and Agriculture Organization, most marine fishing areas have already "reached their maximum potential" for production; further, the majority of the fish stocks are "fully exploited" (FAO 2000). The FAO does not expect that there will be any further growth in the annual catches. Yet over one-fifth of the global population rely on fish as their main source of protein (Ghazi, Smith, and Trevena 1995). As the population increases, there will be more demands for inexpensive food sources such as fish.

Many factors affect the fish stock. One of the problems is that as fishing areas close to coastlines are depleted, people have to travel farther to find a reasonable catch. These travels often take fishermen beyond their territories and into waters that other states believe are within their jurisdiction. In addition, fishing companies have created new technologies to increase catch size. They use drift nets that allow them to "fish" over a 20–40 mile span. These nets do not discriminate between the target fish (the species desired) and "bycatch" (fish that are captured in the net but usually discarded). Further, many companies employ factory ships that allow them to more efficiently process the fish at sea. The combination of these new technologies that increase catch and the growing demand for ocean resources has led to several conflicts over fishing rights.

The United Nations (UN) led the way in resolving these conflicts by initiating treaty negotiations that would help manage ocean resources. The UN Conference on the Law of the Sea involved a series of meetings, from 1973 to 1982, and culminated in the signing of the Law of the Sea Treaty (LOS). One of the major results of LOS was that territorial waters were given political definitions. According to LOS, a state has jurisdiction over its territorial waters, which are defined to be 12 miles from the coastline; this is an area in which the state has control over the resources and that the state may defend as part of its territory. Beyond that point, up to 200 miles, the area is considered an exclusive economic zone (EEZ); this area is not part of the sovereign state, but the state has the exclusive right to control ocean resources in the area. This means that the state can use all of the resources found within the EEZ or sell rights to other entities (states or companies) to use the resources. Anything beyond the EEZ is considered the global commons area, an area that is to be shared by all peoples and managed by the international community.

Though there are still many problems with managing the resources in the global commons area and resource depletion is a major concern, it was envisioned that LOS would prevent most of the conflict over fishing rights. Coastal states would no longer be threatened by foreign fishing ships right off their coasts; the limits of the coastal states' jurisdiction were now clearly determined; and the coastal states had the opportunity to sell rights to use the resources in the EEZ. It was believed that through multilateral cooperation, violent conflicts over this limited resource would be avoided.

Fish wars: the fight for turbot. Violence grew imminent in 1995 when Canada and the European Union (EU) fought over fishing rights in the Grand Banks, more than 200 miles off the coast of Newfoundland, Canada. Even though the fishing grounds were beyond its EEZ, Canada believed it had the jurisdiction to protect the "straddling" stock of turbot. A straddling stock consists of fish that literally straddle a state's EEZ and the global commons area. According to international law, a state has jurisdiction over the stock when it is in the EEZ but does not have jurisdiction when the stock is in the high seas. The Northwest Atlantic Fisheries Organizations (NAFO), an international governmental organization created to manage the fishery, set quotas for 1995: 3,400 tons for EU boats and 16,300 tons for Canadian boats. Canada charged that trawlers from both Spain and Portugal (both members of the EU) had been fishing over the NAFO-set quotas for the previous five years. Someone needed to take action to protect the stock, and Canada believed it had the right to do so (DeMont 1995a); David Ralph Matthews refers to this as "the moral regulation of nature" (1996). On March 3, 1995, Canada declared a sixty-day moratorium against turbot fishing while the parties negotiated an agreement over the quotas. When Spain and Portugal refused to honor the moratorium, the situation grew tense.

The Canadian coast guard attempted to board one of the Spanish fishing trawlers, the *Estai*. When its attempts were evaded, Canada fired warning shots over the bow of the *Estai*, seized it, and jailed the crew. Canada also cut the nets of several other boats. When the Canadians examined the nets, they found that the mesh was too small and that only 2 percent of the turbot catch had reached spawning age; both situations violated international agreements (Russell 1995). EU officials threatened to send warships. Spain threatened to take the case to the International Court of Justice (ICJ) and sent several navy patrol vessels to the Grand Banks area.

After many weeks of threats and negotiations, an agreement was finally reached. Canada gave up some of its catch quota to the EU in exchange for the implementation of tough surveillance and enforcement measures, including onboard inspections and satellite monitoring (DeMont 1995b). However, the negotiated cease-fire was short lived when Spain decided to pursue its interests in the ICJ. Spain argued that Canada had no authority to take action against Spanish vessels that were in the high seas and that the seizure of the *Estai* was illegal. Canada's argument was relatively simple. On May 10, 1994, Canada had placed a reservation on its acceptance of compulsory jurisdiction of the ICJ. (In order for the ICJ to hear a case, each state must formally accept the jurisdiction of authority of the International Court of Justice. When they accept this compulsory jurisdiction [or at some later time], they may place a "reservation" on their acceptance; this means that they may state the circumstances under which their acceptance of the court's jurisdiction would not apply.) The reservation said that Canada would not accept the authority of the ICJ on conservation and management issues with respect to the NAFO area. In other words, Canada withdrew its acceptance of the court on this, and any other related, issue. Before the ICJ can decide the merits of a case, it must determine if it has the authority (jurisdiction) to hear the case. In a twelve to five decision the ICJ had little choice but to rule that it did not have jurisdiction to hear this case (ICJ 1998). Spain, in essence, lost.

Was ongoing violence averted in this case? Yes. One can make the argument that the ICJ decision was an important win for Canada, but the court left the issue of appropriate management of straddling stocks in dispute. We may not have heard the last of this issue.

■ Nonrenewable, Transboundary Resource: Water

The Quran states that water is the source of all life; but water is the primary limiting resource in an arid region. Some experts believe that the next protracted war in the Middle East, one of the most arid regions in the world, will be fought over water, not oil (Gleick 1994; Postel 1993). The situation around the world has become so serious that the United Nations now refers to "water refugees"—the 25 million people who have faced social and economic devastation as a result of the critical shortages of water. Water refugees now outnumber war refugees ("Nor Any Drop to Drink" 2001). Thus, the UN has declared 2003 the International Year of Freshwater.

Source: Danziger, *The Christian Science Monitor,* March 29, 1995. © Copyright 1995 *The Christian Science Monitor.*

States in the Middle East depend almost entirely on river systems for their water supply. The problem is that none of the rivers exist solely within one state's borders; 90 percent of all potable (drinkable) water sources are transboundary (Peterson 2000). Generally, countries that are in control of the water supply are in the most powerful position, since they can control both the quantity of water (through dams) and the quality of water (by their industrial and agricultural actions). This creates a power imbalance between upstream and downstream states, which can lead to either cooperation or conflict.

One of the debates concerns the issue of sovereignty. A guiding principle in international law states that a government has jurisdiction over domestic issues, including natural resources. However, the world is becoming more interdependent each day. The boundary between domestic and international issues is rarely clear, as has been noted in many chapters in this book.

Proponents of absolute sovereignty rely on the Harmon doctrine: A state has the right to make all decisions about the resources that lie within that state's territory. The growing awareness of transboundary resource problems has led to the development of an alternative doctrine—equitable utilization—promoted by the United Nations: States have the obligation to ensure that their use of those resources will not adversely affect other states

(Ahmed 1994). There is still great debate as to which principle should prevail in resource issues. This is particularly true in a region, like the Middle East, that is highly dependent on a transboundary resource for its very survival.

The Tigris-Euphrates river system. The relationship among Turkey, Syria, and Iraq over the Tigris-Euphrates river system is a good example of the typical upstream-downstream relationship that is characterized by both conflict and cooperation. These rivers begin in the Turkish mountains, diverge before they reach Syria, and then join again further downstream in Iraq. Approximately 40 percent of the water resides within Turkish territory, while 25 percent is in Syria and 35 percent is in Iraq (Hillel 1994). All three states are in the process of developing, and their success in this area will be dependent in large part on their ability to maintain a consistent and adequate water supply. The ability to irrigate is the primary issue (Scheumann 1998). Yet, only Turkey has adequate rainfall to meet its needs (Drake 2000).

Turkey is the source of approximately 70 percent of the Tigris-Euphrates river flow (Drake 2000). It has worked hard to develop its water resource with an elaborate dam system (the Southeast Anatolia Project) to create storage, generate hydroelectric power, and prevent floods. Turkey filled its fourth dam on the Euphrates river in 1999 and has begun building its first dam on the Tigris. When completed, the project will include twenty-one to twenty-two dams and seventeen hydroelectric power plants and will reduce the flow of the Euphrates by 30–50 percent (Drake 2000; Hottelet 1996; Hillel 1994). Turkey's goal is to become self-sufficient in food production, and it believes that its water development projects are critical to meeting this goal. A statement by Turkish prime minister Suleyman Demirel in 1992 clearly indicates Turkey's commitment to the Harmon principle: "We do not say we should share their [Syria and Iraq's] oil resources. They cannot say they should share our water resources" (Postel 1993: 16). The potential for conflict is great.

Much of Syria's difficulties with water can be found in two areas: poor relations with its neighbors and poor policy decisions. Syria has access to both the Tigris-Euphrates and the Sea of Galilee. However, it is reliant on goodwill with its neighbors (Turkey and Israel) to secure its water rights. Turkish-Syrian relations are "the worst ever" (Hottelet 1996). Turkey has on many occasions stopped the flow of water to either fill a new dam or repair existing dams—an action that, in Syria's eyes, is an assertion of its power over the water issue. Relations with Israel are better, but a peace treaty is delayed over water rights to the Sea of Galilee. Since water is such a scarce resource, access to the Sea of Galilee is not only a symbol of power but a requirement for survival ("Peace" 2000).

The other problem for Syria is poor policymaking. Syria has developed the Euphrates, but this, like other Syrian projects, has not been successful.

The Tabqa Dam was built in 1974 but failed to reach its potential for both irrigation and hydroelectric development. Syria has invested an enormous amount of its gross domestic product (GDP) on security issues, leaving fewer resources available to develop the Euphrates. Further, Syrian strategy includes the overt policy to increase population. Numerical growth is perceived by Syria to be the key to future power and success in the Middle East. Yet Syria has become a net food importer and has so mismanaged its water supply that the groundwater is also depleted (Hillel 1994). Evidence suggests that Syria misuses much of its water supply from Turkey, through overirrigation. Resulting evaporation from Lake Assad (a reservoir) has led to local desertification ("Sharing" 1999). Thus, the flow from Turkey remains critical. In addition, Syria must be concerned about the consequences of interrupting the flow into Iraq.

By some accounts, Iraq has the most to lose in this transboundary issue. Iraq must absorb losses from both Turkish and Syrian water projects; this is particularly critical since Iraq is already the most arid of the three countries. Further, the water quality is diminished greatly by the time it reaches Iraq, with high concentrations of agricultural chemicals and salts derived from Turkish and Syrian irrigation (Gleick 1994). Complicating this situation is that UN sanctions from the Gulf War have left Iraq in serious economic trouble. There is insufficient money to repair irrigation channels and water pumps; thus, fields have been left so dehydrated that only a thick coat of salt remains ("Sharing" 1999). Further, Iraq relies on expensive chemicals to purify its water for drinking, but UN sanctions prevent Iraq from importing those chemicals. The UN Children's Fund (UNICEF) estimates that only 5 percent of the current water supply is potable, which has resulted in an increase in both diarrhea and respiratory diseases (Nagy 2001). While Iraq strongly believes it has historical rights to both the Euphrates and Tigris rivers, it recognizes that it is in a poor negotiating position since it is the most downstream. If Turkey and Syria complete their water projects as planned, Iraq's share of the Euphrates could be reduced as much as 80 percent (Gleick 1994). Once UN sanctions are lifted, Iraq will be able to buy food and supplies with its oil income; however, Iraq is still concerned about ensuring its own water security.

There is certainly a great need and potential for cooperation in this case. If the three states follow through with their water projects for both the Tigris and Euphrates, they will exceed the total annual flow of the rivers. Since dams and diversions also increase the rate at which the water evaporates, the states, with their individual water projects, will also reduce the total amount of the resource (Scheumann 1998; Hillel 1994). Yet there has been limited success in cooperation. Turkey and Syria concluded an agreement in 1987 that guaranteed Syria a set flow from the Euphrates. Syria, in turn, concluded an agreement three years later that guaranteed Iraq a set flow. The cooperation was short lived.

In 1990, Turkey began to divert water from the Euphrates to fill the Ataturk Dam, the key point of the Southeastern Anatolia Project. For one month, Turkey completely shut off the Euphrates, sending no water downstream to Syria and Iraq, although Turkey sent additional water for several months preceding the cutoff. Both Syria and Iraq protested to the Turkish government and threatened military action; Turkey resumed release from the Euphrates. While violence was avoided in this instance, threats of using the water "weapon" to gain political leverage have continued. Turkey threatened to cut off the water supply to Syria in 1989 because of its support for the Kurds, a group of indigenous people who have been persecuted by many Middle Eastern countries (Postel 1993). During the crisis in the Persian Gulf, the United Nations considered using Turkish control over the Euphrates as a way to force Iraq to withdraw from Kuwait (Gleick 1994).

Cooperation on this issue is complicated by several things: existing political tensions in the Middle East region, disagreement over the amount of water that actually exists in the Tigris-Euphrates river system, disagreement over the parties' need for the water, concerns about water quality, and the possibilities of water from other sources (Scheumann 1998).

The predominance of animosity and conflict in this case has had one rather interesting and unexpected side benefit: The World Bank has stated that it will not support further water projects until the three states reach a mutually acceptable agreement to share the water resources (Hillel 1994). An additional concern has come from the British government; it has reservations about providing credit guarantees for the Turkish dam project on the Tigris because of the threat the dam poses to the Kurdish community that lives there ("Turkey's Latest" 2000). These actions by both the World Bank and the British government are providing outside incentive to protect the Tigris-Euphrates system from further alteration, and they have alleviated the immediate pressure for conflict. The future, however, is not clear.

◼ Nonrenewable, Boundary Resource: Oil

Technically, oil is a renewable resource; it is created when dead marine microorganisms accumulate on the ocean floor and eventually become released as hydrocarbon molecules. The key word is *eventually;* fossil fuels are called such simply because it takes thousands of years for supplies to become abundant. In fact, some estimate that humans now extract each year what it has taken nature 1 million years to create (Pickering and Owen 1994). This is the problem: Oil reserves are a limited and finite resource, and industrialized states use more of this resource than they can produce themselves. Currently, the United States is the largest user of petroleum and uses about 25 percent of the global production (NRDC 2001), approximately 19.4 million barrels each day (MacKenzie 2001). However, the largest oil fields in the United States contain only 3–4 percent of the global reserves (NRDC 2001), so it imports a significant amount of oil from the Middle East.

Two issues must be considered when discussing the global oil situation: consumption requirements and domestic availability. Oil consumption is directly tied to development and industrialization. Most major factories and industries rely on oil, as does agribusiness, which relies on mechanized farm equipment. A state's desire to maintain a strong and robust military also increases the need for oil, as do citizens' desires for convenient transportation. These development needs have increased at a phenomenal rate since World War II. The United States was able to accommodate its needs for many decades; in fact, prior to World War II, the United States was the primary producer of oil (Võ 1994). However, the needs soon outpaced the production capability and the U.S. reserves.

Approximately 65–76 percent of the global oil reserves are in the Middle East (USDOE 2001). Thus, the Middle East has become the site of an important commercial enterprise that affects both economic development and international politics. The United States has become quite reliant on Middle Eastern oil; despite its attempt to limit oil imports during the Organization of Petroleum Exporting Countries (OPEC) crisis in the 1970s, overall oil consumption has risen by 12 percent since 1973, and it has risen 42 percent in the transportation sector (MacKenzie 2001). The United States now imports more Middle Eastern oil than it did before the crisis. Other developed states also depend on imported oil. Japan has extremely limited reserves of fossil fuels and must rely almost completely on imports, most from the Middle East (Pickering and Owen 1994). The Europeans are less reliant on oil than the United States, but their reserves are also much smaller and more expensive to exploit (Flavin 1991). They also consume more than they produce, and they are heavily dependent on the Middle East for their supply.

As the situation stands, there are many powerful states in the North that depend heavily on a resource that is necessary for industrial development, personal consumption, and maintenance of the military. One could make an argument that the power in these states is determined in large part by oil. But that resource is very limited and is predominantly found in another region of the world. While this sets up the potential for cooperation, it has, in fact, more often led to conflict.

The 1991 Gulf War. On August 2, 1990, Iraqi military forces crossed over the border and seized the sovereign state of Kuwait. While the Iraqi government made many claims to justify the action, one of the major issues was oil. The Rumailla oil fields are on the border between the two states. Iraq claimed that because Kuwait was using too much of that oil reserve and was stealing oil from the Iraqi side of the border, Iraq should be compensated (Freedman and Karsh 1993). This dispute has its roots in history.

Source: Ed Stein, *Rocky Mountain News,* 1990. Used by permission of Ed Stein.

In the nineteenth century, Kuwait was a province of the Kingdom of Iraq. When the Ottoman Empire, which controlled Iraq, ended, Kuwait became a protectorate under Britain and remained so until its independence in 1961. It was later discovered that some of the richest oil reserves were within Kuwaiti territory. Development in Kuwait grew tremendously throughout the 1960s, so that by the 1970s, Kuwait had one of the highest per capita GDPs in the world (Võ 1994). Although the Kuwaiti economy slipped during the 1980s due to inflationary problems, Kuwait was clearly a state that was very well off economically.

While Kuwait was growing at unbelievable rates, Iraq faced its own internal problems. The growth of the Kuwaiti economy was a sore point to Iraq. Iraq still believed that it had historical rights to the territory and had threatened to fight Britain during its protectorate period (Freedman and Karsh 1993). The Iraqi economy faced an even greater strain during the 1980s: the war with Iran. The eight-year war left an enormous drain on the Iraqi economy, which was made worse by the fact that Kuwait supported and lent money to Iran. The economic strain, coupled with disputes over Iraqi rights to Kuwaiti territory, paved the way for the Iraqi invasion in 1990.

The invasion of another sovereign state was a clear violation of international law, and the United Nations responded. Yet many other clear violations have been ignored by the international community. What made this so different? Many experts suggest that the only reason the United States and the rest of the North took such swift and decisive military action was their concern for protecting the oil supply (Võ 1994; Pickering and Owen 1994). The North, especially the United States, had many concerns. It was worried not only that the Kuwaiti oil reserve would be in Iraqi hands but that Iraq might destroy the oil fields (Warner 1991). Further, there was great concern that Iraq would not stop with Kuwait but would attempt to claim the abundant oil reserves on Saudi territory as well. No matter what the goal of the Iraqis, the North felt that its supply of petroleum was in danger.

What followed was an example of collective security provisions enacted in the United Nations. The UN Security Council condemned the invasion and demanded that Iraq retreat from the borders of Kuwait. When Iraq refused, sanctions were imposed and a multinational military force was created. A short but violent war ensued, with the final result being the Iraqi withdrawal from Kuwait. Although Iraq did destroy some of the oil fields and set as many as 500 oil wells on fire (Warner 1991), the violence produced the desired result: protection of the oil reserves. However, the story did not end in 1991.

As of 2002, the United States continues to keep 4,500 troops stationed in Kuwait, and another 12,000–19,000 troops throughout the Middle East *just in case* Iraq should threaten the oil supply again. These troops regularly patrol over the Middle Eastern skies, hold regular training exercises, and maintain the tanks and other supplies needed *in case* there is another conflict. In addition, the U.S. Navy's Fifth Fleet maintains vigil off the coast of Bahrain (Barr 2001). Yet despite these costs, Americans continue to pay $1.30–2.50 per gallon of gasoline, and consumption continues to grow. Environmental economists suggest that if we paid the *true* price for our products—including government subsidies and support, environmental impact, and perhaps human lives—we would begin to recognize the need to conserve. Instead, these costs get absorbed in the general budget, and become, seemingly, lost to the general public. As the cartoon on p. 270 suggests, Americans would probably be more willing to conserve if they had to pay the *true* cost of recovering a scarce natural resource.

■ CONCLUSION: THE NEED FOR SUSTAINABILITY

One of the things that can be learned from the case studies presented in this chapter is that the potential exists for both cooperation and conflict in all natural resource issues. Cooperation is more likely to occur when the parties are otherwise on friendly terms and when the resources are neither severely limited nor critically important. Conflict, particularly violence, is more likely when the parties distrust each other, the resources are both limited and important, and cooperative mechanisms are insufficient.

Perhaps a more important lesson to learn is that there is a great need to alter our use of natural resources to a more sustainable level so that cooperation can be encouraged and conflict avoided. As discussed in earlier chapters, the UN World Commission on Environment and Development has defined *sustainable development* as "development which meets the needs of the present without compromising the ability of future generations to meet their own needs" (Elliott 1994). The goal is to ensure that while basic human needs are being met now, we do not jeopardize the ability of future generations to meet their needs. For the South, sustainable development means finding alternatives to many of the development techniques that are currently depleting resources. This might result in finding inherent value in standing forests—rather than valuing deforested plots of land—or developing effective fishery management plans so that stocks are not expended.

For the North, sustainable development means reducing consumption in general. This will require developing plans for more efficient use of current resources as well as exploring more fully solar, wind, and thermal energy. Both North and South can cooperate to develop policies that are more efficient and that ensure the provision of basic human needs around the world.

While natural resource issues have been perceived by many policymakers as "low politics," it is clear that natural resources can become "high politics" if the stakes are high enough and the needs critical enough. One solution to avoid conflict and encourage cooperation in natural resource issues is to live more sustainably. This may be, in fact, the primary challenge for the twenty-first century.

■ DISCUSSION QUESTIONS

1. Use the fourfold matrix (Table 14.1) to determine in which resource areas cooperation and conflict are most likely to occur.
2. Are there other resources in which the debt-for-nature model might be useful?
3. Should the price of gasoline accurately reflect its cost?
4. If you were president of the United States and faced a threat to Middle Eastern oil reserves similar to the one George Bush faced in 1990, what would you do?
5. The United States contains only 5 percent of the global population but uses 25–40 percent of global resources. Does the United States have a responsibility to try to reduce its national consumption levels? What could it do to accomplish this?

■ SUGGESTED READINGS

Drake, Christine (2000) "Water Resource Conflicts in the Middle East." *World & I* 15, no. 9 (September): 298+.

Klinger, Janeen (1994) "Debt-for-Nature Swaps and the Limits to International Cooperation on Behalf of the Environment," *Environmental Politics* (summer).

Matthews, David Ralph (1996) "Mere Anarchy? Canada's 'Turbot War' as the Moral Regulation of Nature." *The Canadian Journal of Sociology* 21, no. 4 (fall): 505–522.

Pickering, Kevin T., and Lewis A. Owen (1997) *An Introduction to Global Environmental Issues,* Second edition. London: Routledge.

Russell, Dick (1995) "High-Seas Fishing: Lawless No Longer," *Amicus Journal* 17, no. 3.

Scheumann, W., and M. Schiffler, eds. (1998) *Water in the Middle East: Potential for Conflicts and Prospects for Cooperation.* Berlin: Springer.

Võ, X. H. (1994) *Oil, the Persian Gulf States, and the United States.* Westport, CT: Praeger.

15

PROTECTING THE
GLOBAL COMMONS:
SUSTAINABLE DEVELOPMENT IN
THE TWENTY-FIRST CENTURY

Marion A. L. Miller

Development strategies in the twenty-first century will be shaped by the increasing tensions between economic health and ecological health. Capitalism, the predominant economic system, depends on repetitive expansion. This process of expansion needs and supports industrial processes with insatiable appetites for resources such as oil, coal, wood, and water, and it depends on increasing use of land, sea, and atmosphere as sinks for the deposit of wastes. Clearly, there is a conflict between the economic system's implacable demand and our limited and shrinking supplies. Although there is concern that the ravenous economy will gobble up and despoil the global commons of the atmosphere, land, and oceans, some scholars and policymakers suggest that this catastrophe can be avoided: With sustainable development, we can have economic growth while protecting the global commons (WCED 1987b; UN 1992).

The term *commons* can be used to refer to any of a varied group of resources, including natural resources such as land, forests, atmosphere, oceans, and biodiversity. For these natural resource commons, exclusion is difficult or impossible, and each user has the ability to subtract from the welfare of the other users. For a commons such as the atmosphere, exclusion is impossible. Commons can also include resources such as land and forests, where exclusion is possible, but even in these cases excluding other users does not necessarily mean excluding their impact, since uses in adjoining, nearby, or even distant areas can have environmental impacts on the enclosed area. In spite of the artificial boundaries that might be imposed, these natural resource commons are global commons, and action by any country can have implications for other countries. Consequently, sustainable development at a country level can only succeed if sustainable development is also pursued as a global strategy.

Sustainable development has a multiplicity of definitions. Generally, it implies that it is possible to achieve sound environmental planning without sacrificing economic and social improvement (Redclift 1987: 33). Some definitions emphasize sustainability, and therefore the focus is on the protection and conservation of living and nonliving resources. Other definitions focus on development, targeting changes in technology as a way to enhance growth and development. Still others insist that sustainable development is a contradiction in terms, since development as it is now practiced is essentially unsustainable. The World Commission on Environment and Development (commonly knows as the Brundtland Commission) underlined concern for future generations by asserting that sustainable development is development that "meets the needs of the present without compromising the ability of future generations to meet their own needs" (WCED 1987a: 8). Definitions of sustainable development tend to focus on the well-being of humans, with little explicit attention to the well-being of nature. However, the World Conservation Union (IUCN), the United Nations Environment Programme (UNEP), and the World Wide Fund for Nature (also known as the World Wildlife Fund, or WWF) have proposed a definition that includes nature and highlights the constraints of the biosphere: Sustainable development is "improving the quality of human life while living within the carrying capacity of supporting ecosystems" (IUCN et al. 1991: 10). Although this definition observes the traditional hierarchy that places human beings above the natural world, it does emphasize our dependence on the biospheric envelope in which we live.

■ THE TRAGEDY OF THE COMMONS
   AND ENVIRONMENTAL POLICY

If development is to be genuinely sustainable, policymakers will have to make substantial modifications to their strategies and their assumptions. They have long urged regulation and private ownership (privatization) as management strategies for the commons. This perspective has been informed by the tragedy of the commons paradigm. Garrett Hardin (1968) illustrates the problems attendant on the management of the commons by using the example of a common pasture. Herders are free to add livestock to the pasture. This arrangement is satisfactory, but, after a time, the carrying capacity of the land is reached. If any more livestock is added, the pasture will begin to be degraded. But each herder continues to act in his own self-interest. The inevitable result of unrestrained individual action is overgrazing of the common pasture. Although each actor is aware of the consequences of his action, he feels that individual restraint would hurt him disproportionately. If he exercised restraint, he would still share in the costs resulting from the degraded commons, while being excluded from the short-term or medium-term benefits that could be gained from the addition of more animals to the pasture. He regards the cost to him of restraint as being more significant than the cost of the depletion of the common pasture; consequently, he would prefer to join with the other herders in their continued exploitation, even if this leads to the complete degradation of the pasture. With every herder acting in his own self-interest, the inevitable outcome is competitive overexploitation of the commons. This freedom to satisfy self-interest results in a tragedy of the commons.

Hardin (1968) extrapolated from the example of the common pasture to other commons resources and warned of emerging tragedies in areas such as population and pollution. He felt that social and political controls were needed, with some commons reorganized as private property and others kept as public property. Private property is seen as an effective solution, because it is assumed that people and institutions will recognize that it is in their interest to protect any property that they own. Public property would be managed by the state, and the state would allocate the right to use it. Essentially, Hardin encouraged the restriction of freedom and the enclosure of resources. Enclosure includes private ownership by individuals and institutions as well as states' rights to control access.

Some scholars have questioned both the assumptions and the implications of Hardin's model. They argue that his assertion that overexploitation is inevitable in the absence of privatization or government controls is much too sweeping. They charge that his diagnosis is faulty because his model assumes that commons are always readily accessible to all. In reality, natural resource commons are held in a variety of property arrangements. These include open access, private property, communal property, and state property arrangements. Global commons such as the oceans and the atmosphere are clearly open access commons. Large tracts of commons resources are held as private property. Other commons are the exclusive property of a specific community. Finally, some commons, such as land and marine parks, are state property, and the state controls access to these (Berkes et al. 1989: 91–93). The critics maintain that although Hardin foresees a tragedy of the commons, his diagnosis would have been better termed "tragedy of open access" since it was specifically applicable to commons with an open access property arrangement (Stevenson 1991: 3) The other types of property arrangements have built into them regulation of access to the commons. While they do not guarantee an avoidance of degradation, they still do not reflect Hardin's example.

Hardin's perspective to the contrary, narrowly defined self-interest will not always be the determining factor. Users can regulate themselves, and the commons do not have to be despoiled by the drive for individual gain. For example, scholars have studied successful cases of resource management by communities. In these cases, the resources are regarded as the common property of specific communities that have formal or informal rules regarding resource use and management (see Berkes et al. 1989; Ostrom 1990; Stevenson 1991). The emphasis is on community welfare rather than naked self-interest.

■ PROTECTING THE GLOBAL COMMONS

Whether or not the tragedy of the commons is an appropriate paradigm for explaining the exhaustion of resources and environmental degradation, the global environment is under threat. This threat has been salient enough to make the environment an item on the international agenda. In the period between 1900 and 1968 there were a series of bilateral and multilateral agreements and conferences on environmental issues, many of them focusing on particular species or regions. For the most part, they had an optimistic tone, taking the perspective that the existing problems could be fixed by the employment of financial and technological resources. But by the time of the Biosphere Conference (formally, the Intergovernmental Conference of Experts on the Scientific Basis for Rational Use and Conservation of the Resources of the Biosphere) in 1968, that optimism had been dimmed, and conference participants were emphasizing that environmental deterioration had reached a critical threshold (Caldwell 1990: 27). This sense of crisis was also to be reflected in the environmental conferences held in 1972 and 1992.

■ The 1972 Stockholm Conference

The Stockholm Conference (formally, the United Nations Conference on the Human Environment) was not the first environmental conference to draw representatives from many of the world's nations. But it was the first large-scale environmental conference to look beyond scientific issues to broader political, social, and economic issues. This 1972 conference put the environment squarely on the official international agenda. Delegates came from 113 nations, twenty-one United Nations agencies, and sixteen intergovernmental organizations (UN 1972). In addition, more than 200 nongovernmental organizations (NGOs) sent observers.

Since the conference took place at a time when issues of global equity were becoming more prominent in international forums, it is not surprising that this was reflected in preconference discussions and at the conference. Developed countries and developing countries clashed over global economic relations and environmental politics. Before the conference, developed countries had identified a particular set of issues to be addressed at the conference—issues such as pollution, population explosion, conservation of resources, and limits to growth. But the developing countries wanted to enlarge the agenda to include issues such as shelter, food, and water. They were able to use their voting power in the General Assembly to press developed countries to adopt a more inclusive agenda (McCormick 1989: 92–105).

The most significant institutional outcome of the Stockholm Conference was the creation of the United Nations Environment Programme (UNEP). In the decades since Stockholm, UNEP has played a significant role in shaping the environmental policy agenda and in coordinating environmental policy. The conference output also included the Declaration on the Human Environment, the Declaration of Principles, and Recommendations for Action. One of the minor resolutions addressed plans for a Second United Nations Conference on the Human Environment.

The 1992 Earth Summit

The second conference was held on the twentieth anniversary of the Stockholm Conference. Officially designated the United Nations Conference on Environment and Development, it was popularly termed the Rio Conference or the Earth Summit. It attracted greater official and unofficial interest than had the Stockholm Conference. More than 170 nations sent delegates, and 118 of these nations were represented by their heads of state. Thousands of NGOs sent representatives, and there were nearly 9,000 members of the media (WRI 1993: 9).

The conference and the preparatory work that preceded it showed that there were still significant differences between the developed and the developing world. Consequently, each group provided different contributions to the agenda: Developed countries wanted to focus on ozone depletion, global warming, acid rain, and deforestation, while developing countries also wanted to explore the relationship between the economic policies of the developed countries and the sluggish economic growth of developing countries. The concern was that an "environmentally healthy planet was impossible in a world that contained significant inequities" (Miller 1995: 9).

The output of the Earth Summit included the Rio Declaration, new treaties on climate and biodiversity, a statement on forests, and Agenda 21, which is an action plan for sustainable development. Additionally, conference participants agreed that they should begin preparatory work for a treaty to curb desertification. They also set up a watchdog commission to monitor progress on the commitments made at Rio.

▨ *The Rio+5 Conference*

The Rio+5 Conference (also called Earth Summit II) was held in June 1997. It afforded delegates the opportunity to assess the progress that had been made since the 1992 conference and to plan for continued implementation of the Earth Summit strategies.

Government officials made the usual grand speeches and fulsome promises, but analysts and NGOs were critical of the progress made at the 1997 conference and in the preceding five-year period. Gordon Shepherd, WWF's director for International Policy, criticized the conference for spending too much time on speeches and not enough time establishing the targets and timetables necessary to implement commitments made in 1992. He criticized the lack of progress on issues such as forests, climate, and toxic chemicals. He noted that the conference should have taken the initiative to ensure that international economic agreements were supportive of sustainable development. Shepherd did identify one area in which progress had been made: Agenda 21 was being implemented through community and local action groups, small enterprises, and some governments (WWF 1997).

Martin Khor, director of the Third World Network, expressed similar opinions. He was disappointed at the disappearance of the spirit of Rio, demonstrated by the fall in aid to developing countries, the continued drain of resources from developing countries, and the new intellectual property rights agreement of the World Trade Organization (WTO). Meanwhile, however, he was encouraged by the actions of local community leaders and millions of ordinary people around the world (Khor 1997).

▪ ECONOMIC POLICY AND THE GLOBAL COMMONS

At the same time that environmental policy initiatives are being developed to address environmental concerns, economic policy initiatives are being made that have the potential for environmental disaster. An examination of the World Trade Organization illustrates some of these concerns. The WTO is one of the results of the Uruguay Round of the General Agreement on Tariffs and Trade. The Uruguay Round had done much to create an environment that will enhance transnational corporations' ability to extract profits. The goal of the WTO is the harmonization of international standards for trade in goods and services, as well as for intellectual property rights. Such harmonization will mean that corporations will find fewer regulatory obstacles to their operations in host countries. Government regulations that require that imported products meet certain standards with regards to content or process are subject to challenge. As a result, environmental standards have been subject to challenge. Conservation measures, too, can be challenged. For example, a country might impose measures to restrict the exports of forestry resources or fish, but this could be ruled an unfair trade practice. If a local corporation can buy the resource, that right also has to be extended to foreign-based corporations. Preferential treatment cannot be given to local interests. If the WTO panel rules against a country, it has to make the recommended change within a prescribed time or face financial penalties or trade sanctions. These changes in trade rules have clear implications for sustainable development. Local jurisdictions lose access to some of the regulatory tools that might be used to shape development.

The WTO harmonization of intellectual property rights is another change with differential impacts: It supports the interests of transnational corporations, but it imposes costs on some local communities and on the environment. Nation-states have a variety of patent regulations, but the WTO's rules seek to harmonize intellectual property rights among all its member states. These rules are presented in the WTO's Trade-Related Aspects of Intellectual Property Rights (TRIPs) agreement (Agreement 1994), a comprehensive multilateral agreement addressing several areas of intellectual property, including patents.

Patents give investors temporary monopoly rights. They can exclude others from making and using the invention for a limited time. This right of exclusion underscores the conflict between private gain and public interest, as transnational corporations appropriate common knowledge and the fruit of many centuries of development. The TRIPs agreement will require member countries to extend patent protection to products as well as processes (see Article 27[2]). As a result, anyone using new processes to develop patented products would still be required to pay fees to the holders of the patents (Prashad 1994). Corporations based in the major industrialized countries favor product-based over process-based patenting laws. They can patent a product with a technologically costly production process without being concerned that they would suffer significant financial harm as a result of the development of less costly or more environmentally benign production processes. If a less costly process is developed, fees still have to be paid to those who hold the patent for the product. A product patent therefore discourages the kind of innovation that would seek lower-cost, labor-intensive methods and would be suited to the resources of many third world countries. If a U.S.-based drug company has a valid patent for a particular vaccine, and another institution (say, in Nigeria or Brazil) develops a vaccine for the same disease using a different process and lower-cost inputs, the second product could be challenged for infringing on the original patent.

Under the TRIPs agreement, patent protection will expire twenty years from the filing date. The agreement is already in effect in the industrialized countries, but third world states were granted a ten-year transition period in which to extend intellectual property rights to sectors not previously subject to intellectual property protection (see Article 66[1]).

The WTO, then, is continuing the process of enclosure that is essential to the health of the economic system. It has a relentless, insatiable appetite for more raw material resources, processes, and markets. As the capitalist world-economy broadens or deepens, more areas become enclosed or incorporated into the economic system. Enclosure is essential for the survival of capitalism because capitalism depends on the institution of private property. Private property allows the owners of capital the right of exclusion (Heilbroner 1985: 38). Since they can legally prevent others from using their property, they have the exclusive right to use particular resources to extract profit from the system. Clearly, this creates an incentive for individuals or institutions to enclose commons (Miller 2001: 113)

The enclosure of natural resources includes not just land and water but also living organisms. The WTO is a critical forum for establishing the principles and legal framework for patenting biological materials and life forms. The TRIPs clause addressing living organisms is ambiguous: Article 27(3b) states that parties may exclude from patentability

> plants and animals other than micro-organisms, and essentially biological processes for the production of plants or animals other than non-biological and microbiological processes. However, Members shall provide for the protection of plant varieties either by patents or by an effective sui generis system or by any combination thereof.

As capitalism broadens it encloses more natural resources, and as it deepens it incorporates other kinds of resources, such as knowledge. As a result, one consequence of this deepening of capitalism has been the commodification of intellectual property. Owners of capital with access to scientific and technological resources have been able to use these resources to gain an advantage in economic competition. Both knowledge of nature and natural processes are being enclosed. This enclosure affects not only "new" knowledge but also knowledge that is centuries old. It is an enclosure with significant implications for biodiversity, in particular, and for sustainable development, in general.

■ BIODIVERSITY AND THE GLOBAL COMMONS

Biological diversity, or biodiversity, refers to the variety of life on earth. An assessment of biodiversity requires an understanding of various environmental issues. Clearly, activity in every other area of environmental concern—climate change, deforestation, ozone depletion, desertification, food security, water scarcity, and the like—has implications for biodiversity. To understand and assess biodiversity, it is necessary to examine three categories:

genetic diversity, species diversity, and ecosystem diversity. Genetic diversity addresses the variations of genes within a species; species diversity refers to the variety of species within a region; and ecosystem diversity describes the number and distribution of ecosystems. Biodiversity is held in a mix of property arrangements. It is spread over a wide array of ecosystems, including oceans, forests, rivers, lakes, deserts, mangroves, and backyards. Consequently, it is held in each of the four basic property arrangements: open access, private property, communal property, and state control. Concerns regarding depletion of genetic diversity, species extinction, and disruptions to a wide range of ecosystems have led to increasing attention to biodiversity by a variety of constituencies, including governments, private citizens, corporations, and environmental nongovernmental organizations. Although some of these interests may have conflicting policy prescriptions for biodiversity, they all regard it as an important resource.

Scientists believe between 10 million and 80 million species of plants and animals actually exist, although the majority of them have still not been identified (Ryan 1992: 9). As more and more species become extinct, there is growing demand within the international community for a system to slow or halt the process of extinction. Arguments in favor of species preservation might be selfless or self-interested. One perspective holds that a species has a right to exist because of its unique nature. Another perspective is informed by self-interest: In an interdependent ecosystem, the extinction of some species diminishes the well-being of the other species, including human beings. For many that provides the motivation for the protection of species and threatened ecosystems such as forests, wetlands, and coastal waters (Miller 1995: 110). Whatever the motivation, increased interest in biodiversity led to the Convention on Biological Diversity (CBD).

Clearly, biodiversity is a commons: A decrease in biodiversity has consequences for all species. But this commons issue has particular distributional characteristics, since the bulk of the world's biological diversity is found in tropical forests, which are located in the third world. Both the common-property and distributional aspects of this issue therefore have had implications for negotiation strategies as well as for the convention's prospects for implementation (Miller 1995: 110).

The series of preparatory meetings leading to the adoption of the CBD drew a wide range of interested national and transnational actors, including nation-states, environmental NGOs, and agricultural and pharmaceutical interests. Work on the CBD would address not just the protection of biodiversity but also the use of biotechnology and issues of biosafety. Biotechnology has implications for a variety of fields, including medicine and agriculture, and research in biotechnology depends on a rich pool of genes.

In their efforts to develop new products, pharmaceutical companies are exploring plants, microbes, algae, and fungi. They are also investigating the medicines used by traditional communities. To gain or maintain a foothold in a multibillion-dollar industry, companies and institutions are collecting and screening plant and other natural material. They often use institutions and independent collectors, including third world institutions such as the Chinese Academy of Sciences, the National Biodiversity Institute of Costa Rica, and the Central Drug Research Institute of India (WRI 1993: 8–13).

Although modern agriculture expects to benefit from biodiversity, it has so far proven to be an enemy of biodiversity. Loss of biological diversity has been spurred by modern agricultural practices. If this trend continues, it can reduce our ability to feed ourselves. With fewer species used for food, a plant disease that targets a particular species will destroy a substantial portion of our food supply. Both Ireland and the United States have been hurt by the consequences of genetic uniformity: In the 1840s, the Irish potato crop was destroyed by blight, and more than 1 million people starved; and in 1970 fungus destroyed 15 percent of the U.S. corn harvest (Hobbelink 1991: 64). Most of us eat from a smaller menu of plant species than did our ancestors two generations ago. One estimate is that more than 3,000 plant species have been used for food throughout human history. But today most of the world's food comes from twenty species. Local crops have been suppressed in many areas, and research and production have focused on a narrow range of crops (Juma 1989: 14). In some parts of Africa, the colonizers discouraged the cultivation of traditional crops like millet and encouraged the production of export crops like coffee and cocoa.

■ The Convention on Biological Diversity

All of the various interested parties sought to shape the negotiation of the Convention on Biological Diversity. It was adopted in May 1992 and entered into force in December 1993. The measures include commitments to do the following: adopt biodiversity strategies and action plans; conserve biodiversity by setting up protected areas; restore degraded habitats; conserve threatened species and ecosystems; preserve and maintain the knowledge, innovations, and practices of local and indigenous communities; and ensure the safe use and application of biotechnology. Some of these commitments are couched in ambiguity; as a result, aspects favored by the more powerful actors are more likely to be implemented, whereas those (such as the maintenance of the knowledge and practices of local and indigenous communities) that are likely to benefit the weaker actors are more likely to be ignored. This is clearly illustrated in the neem case.

■ The Neem Case

An exploration of the neem case offers the opportunity to explore the continuing enclosure of the commons and the potential impact on biodiversity.[1]

It looks at some of the consequences of the commodification of nature, when nature is regarded primarily as a commodity to be bought and sold in the marketplace. This commodification underlines the tensions between sustainable development and the predominant economic system. The economic system favors those with economic and political power, and the interests of the powerful are not necessarily supportive of sustainable development.

As capitalism deepens, corporations have begun the process of establishing ownership rights to local knowledge. In the process, they often benefit not only from centuries of indigenous experimentation but also from local scientific research. During the past decade, patents have been sought and granted regarding uses or products related to a variety of plants cultivated by generations of third world communities. The ownership rights granted with regard to the neem tree (*azadirachta indica*) have generated significant controversy.

For centuries the branches, leaves, and seeds of the neem tree have been used in India for medicine, contraception, toiletries, timber, fuel, and insecticide. Its medicinal applications have included leprosy, diabetes, ulcers, skin disorders, and constipation. For at least fifty years, local scientists have studied the seeds' pesticidal powers. The azadirachtin compound in neem has been recognized as an effective biologically selective insecticide. Neem also contains salanin, another powerful pesticidal compound *(Hindustan Times* 1998; Shiva and Holla-Bhar 1996: 148–149). In the past two decades, transnational corporations have been paying particular attention to its use as a pesticide. In an era with increasing concern about dangerous pesticides, the ecologically friendly neem products seem particularly attractive. As a result U.S. and Japanese corporations have declared their interest in neem by taking out U.S. patents for a variety of neem compounds.

In response, critics raise questions regarding authentic intellectual ownership. Neem, with its well-known and age-old uses and properties, is a part of India's common lore. Consequently, the Indian Central Insecticide Board did not register neem products under the 1968 Insecticides Act (Shiva and Holla-Bhar 1996: 152). Although small and medium-sized Indian businesses have long marketed neem-based commercial products, they were not allowed to acquire proprietary ownership of formulas because until Indian law was changed in 1999, it did not permit the patenting of agricultural and medical products. (This was one of the changes made after India became a party to the Paris Convention and the Patent Cooperation Treaty, effective December 7, 1998.)

During the past two decades, however, non-Indian companies have been securing patents for neem products. W. R. Grace & Company has been a major target of criticism for its role. The company has acquired at least four neem patents, which claim an innovative process based on a modernized extraction method. Moreover, W. R. Grace has claimed that its pesticide product is different from traditional formulations because it is more stable for storage than the older products. Critics question Grace's claims of innovation. They assert that, over the centuries, the development of neem products has involved the use of complex processes. And over the past three decades, Indian scientists have made significant improvements in the storage stability of neem products (Lok Swaasthya n.d.).

Local farmers are already experiencing adverse economic consequences from the patenting of neem products. They have to compete with transnational firms for access to neem seeds (Miller 1995: 111). One result of the increased interest in neem has been an increase in the price of neem seeds from 300 rupees to 8,000 rupees per ton (Shiva n.d.). Continuing price increases may put neem seeds out of the reach of farmers who do not have their own neem trees.

Because the patents enclose local knowledge, corporations have been accused of biocolonialism. Individuals, NGOs, and some governments have challenged the granting of patents. After all, the patented products and technologies use biological resources that have long been identified, developed, and used by generations of farmers and indigenous peoples. While corporations stand to reap large profits, local communities receive no compensation and, in fact, they may bear significant economic costs.

Another concern is that the patenting of neem products may reduce genetic diversity. This loss of biodiversity would be a consequence of monoculture and of the alienation of the local community. As corporations become involved in the growing of neem trees to help secure their supply, the tendency is to focus on a few strains, while other strains are neglected and may be lost. In addition, as the traditional rights of local communities are undermined, the capacity these communities have to conserve biodiversity is reduced. This loss of diversity is likely to result in decreased resilience to pests, disease, and environmental stress (Shiva 1997: 88). In addition, if this environmentally benign pesticide becomes more expensive, farmers may employ cheaper and more hazardous substitutes.

NGOs have mobilized to protect third world indigenous knowledge systems and biological resources from piracy, and they have met with some success. They have filed a series of legal challenges against patents granted on biological products. W. R. Grace and its neem products have been particular targets. In September 1995, more than 200 organizations from thirty-five countries petitioned the U.S. Patent and Trademark Office for the revocation of a patent granted to W. R. Grace for a neem pesticide extract. According to the petitioners, the company has wrongfully claimed ownership of an age-old biological process used by millions of farmers for generations (Khor 1998). Although this petition has not met with success, challengers were successful with an earlier petition. The petition was filed in

June 1995 at the European Patent Office (EPO) against Patent No. 436,257, granted jointly to W. R. Grace and the U.S. Department of Agriculture for a method that extracts neem oil for use as a fungicide. The petition argues that the extraction method is decades-old; in addition, the antifungal effects of neem oil have been known in India for centuries and so it cannot be a "discovery" as the company claims (Khor 1998). The petition was filed by the New Delhi–based Research Foundation for Science, Technology and Ecology, in cooperation with the International Federation of Organic Agriculture Movements and Magda Aelvoet, former Green Member of the European Parliament. In its September 1997 preliminary assessment, the EPO indicated that the neem fungicide was not sufficiently novel (IFOAM 1997). An oral hearing followed in May 2000. After the two-day proceeding, the EPO panel revoked the patent.

Campaigners against the patenting of local knowledge have also been encouraged by the repeal of a patent for turmeric (No. 5,401,504) in 1997. Mississippi Medical had obtained the patent, claiming discovery of turmeric's healing properties. But India's Council of Scientific and Industrial Research challenged the patent, using ancient texts to help prove that the claim of discovery was false. The U.S. Patent Office agreed that the claim of novelty was false, and the patent was revoked (Raj 1998).

In the past decade, India has struggled vainly to balance the interests of farmers and environmentalists on the one hand and those of corporations and the global trade and property regimes on the other. As it has attempted the balancing act, the official government position has changed considerably. India's Patents Act of 1970 did not allow the granting of patents for substances in the fields of agriculture and horticulture or for curing or enhancing human, animal, or plant life. Consequently, patents on pesticides and fungicides were disallowed (Patel 1996: 314), and India has been slow to meet WTO requirements. This was reflected in the delaying action in Parliament in March 1995, when the Upper House forced the government to defer the Patent Amendment Bill. But the last few years have seen radical changes. India has acceded to the Paris Convention and Patent Cooperation Treaty, and effective December 7, 1998, it became a member of the International Patent Cooperation Union. In addition, in March 1999, both houses of Parliament passed a bill amending the Patents Act of 1970 and bringing India's patent regime in line with the demands of the WTO. This bill was retroactive and effective as of January 1, 1995.

Under the terms of the TRIPs agreement India is required to offer patent protection to products related to agriculture, pharmaceuticals, and chemicals. Within the country, there is substantial disagreement over this issue. Some Indian citizens and organizations argue that India should accept only process patents and not product patents. But WTO terms require that the government enact a Product Protection Act, further enclosing local knowledge. Corporations are likely to claim that the items they produce with appropriated knowledge are sufficiently novel to merit patenting. This could significantly limit the options of the local community since they might run the risk of violating those patents when they use similar products (Miller 2001: 127).

The Indian government plans to balance this change with legislation providing for the protection of indigenous communities' traditional systems of medicine and the plants used in such systems. While many view this as a welcome move, a similar legislative initiative by another developing country has been challenged as being in violation of the TRIPs agreement. When Thailand drafted a bill aimed at recognizing and protecting traditional healers and medicinal genetic resources, the United States reacted by warning Thailand that it could be in violation of the TRIPs agreement (*The Ecologist* 1997). With its plans for the plant variety protection and farmers' rights bill, the Indian government is signaling its sensitivity to the thousands of farmers who protested against the change in India's patent laws.

In the fight against the enclosure of community resources, some farmers and scientists have proposed an alternative form of intellectual property—collective intellectual property rights, or the collective patent. The collective patent would regard knowledge as a community product subject to local common rights rather than as a commodity available for expropriation by corporations (Shiva and Holla-Bhar 1996: 157–158).

The CBD recognizes farmers' rights to their knowledge of biodiversity. Essentially, it acknowledges that communities of farmers have the right to benefit from the knowledge acquired from generations of cultivating particular plants and breeding selected livestock, but there is no system set up to ensure these rights. The picture is somewhat different for those people and institutions with the resources and inclination to seek intellectual property rights to products or uses that might—or might not—be new. The CBD's language appears to give the advantage to intellectual property rights. According to Article 16, if patents and intellectual property rights are involved, access and transfer should take place on terms consistent with the protection of those rights.

Enclosure changes knowledge of nature into a resource for national and global production, shifting the balance from a community economy where innovations belong to the group to a market economy based on competition and focused on the quest for profits (Gudeman 1996: 103). With such a shift, control of these resources often goes to actors outside the community, and local expertise and knowledge as well as the intrinsic worth of biodiversity are devalued (Shiva 1997: 67).

The new property rights regime privileges the rights of corporations that function as plant breeders over the rights of farmers who have had a long-term nonproprietary relationship with the genetic material of the fields and forests. As a result, farmers cannot benefit from their comparative advantage in the context of the new agricultural technology (Miller 2001: 130).

■ CONCLUSION

The enclosure of neem in India addresses issues of property, access, and equity. It also demonstrates that biodiversity can be threatened, not only by the enclosure of natural resources but also by the enclosure of the knowledge about those resources. Enclosure of the knowledge commons results in reduced or lost access as well as increased inequity. Under the WTO's intellectual property rights system, the rights of those who own capital supersede the common right of others, even if those with common rights have prior claims.

Enclosure means dispossession, and dispossession has both material and nonmaterial consequences. In material terms, the rising costs of some natural resources can affect people's ability to earn a living. As enclosure makes neem less affordable and accessible, farmers may turn to cheaper pesticides that are not environmentally benign. Furthermore, dispossession and economic insecurity undermine social and cultural life. In India some people regard neem as a sacred tree. This link with the sacred becomes more tenuous as the tree is increasingly regarded primarily as a commodity.

Missing from the definitions of sustainable development examined early in the chapter is the issue of equity—equity not only for those of the present generation who have little or no access to natural resources and economic goods but equity also for people of future generations. Equity is clearly an important prerequisite to sustainable development, since both poverty and affluence have been associated with unsustainable practices. But while the affluent can choose to live sustainably, the poor, who may be landless and food-insecure, might be constrained to make such unsustainable choices as cutting down biodiverse forests to grow food crops. A sustainable strategy will have to be based on an awareness of how the environment is viewed and used in different parts of the world. In many third world countries, earning a livelihood involves the exploitation of natural resources such as soil, forests, and minerals, to a greater extent than is the case in the industrialized world. For the industrialized world, the environmental focus tends to be on conservation and aesthetics; however, closer examination shows that much of the exploitation of natural resources in the third world is for the benefit of corporations based in the industrialized world.

The global economic system depends on endless growth, and that is clearly unsustainable. The governments of industrialized countries support this economic system because with growing economies they can pacify the less affluent members of their societies with a slice of a larger pie, instead

of having to share the existing pie more equitably. Third world governments also want their countries' economies to grow; to that end, many have
embraced the Western model of development. However, the planet cannot
support everyone in such a resource-intensive lifestyle.

In spite of the apparent tension between economic health and ecological health, in the long run economic health is dependent on ecological
health. The economy cannot thrive in the face of the total devastation of our
biospheric envelope. If we accept that fact, then there are only two ways to
resolve the tension between economy and ecology: We can let the tension
continue until the integrity of the ecology and the economy deteriorates; or
we can make the economic, social, and cultural changes that will support
sustainable development at the community, national, and global levels.

■ DISCUSSION QUESTIONS

1. What definition of sustainable development do you prefer? Why?
2. Do you support the increasing enclosure of natural resources?
3. Which do you prefer: enclosure of natural resources through private
   ownership or through government restriction?
4. Do you support the patenting of intellectual property?
5. In the neem case, are you more sympathetic with the local farmers
   or with the pharmaceutical companies? Why?

■ NOTE

1. This section on neem is essentially a paraphrase of Miller 2001, pp. 122–128.

■ SUGGESTED READINGS

Hobbelink, Henk (1991) *Biotechnology and the Future of World Agriculture*. Lon-
   don: Zed Books.
Juma, Calestous (1989) *The Gene Hunters: Biotechnology and the Scramble for
   Seeds*. Princeton, NJ: Princeton University Press.
Khor, Martin (1998) "A Worldwide Fight Against Biopiracy and Patents on Life."
   Available online at the Third World Network website, http://www.twnside.
   org.sg/souths/twn/title/pat-ch.htm (accessed 11/16/98).
Miller, Marian A. L. (1995) *The Third World in Global Environmental Politics*.
   Boulder: Lynne Rienner.
——— (2001) "Tragedy for the Commons: The Enclosure and Commodification of
   Knowledge." In Dimitris Stevis and Valerie J. Assetto, eds., *The International
   Political Economy of the Environment*. Boulder: Lynne Rienner, pp. 111–134.
Redclift, Michael (1987) *Sustainable Development: Exploring the Contradictions*.
   New York: Methuen.

Shiva, Vandana (n.d.) "The Neem Tree: A Case History of Biopiracy." Available on-line at the Third World Network website, http://www.twnside.org.sg/souths/twn/title/pir-ch.htm.

―――― (1997) *Biopiracy: The Plunder of Nature and Knowledge.* Boston: South End Press.

Shiva, Vandana, and Radha Holla-Bhar (1996) "Piracy by Patent: The Case of the Neem Tree." In Jerry Mander and Edward Goldsmith, eds., *The Case Against the Global Economy.* San Francisco: Sierra Club Books, pp. 146–159.

WRI (World Resources Institute) (1993) *Biodiversity Prospecting.* Baltimore: World Resources Institute.

Part 5

CONCLUSION

16

FUTURE PROSPECTS

Michael T. Snarr

One question of interest to people studying global issues is, given the critical nature of these issues, what will the future look like? Will humans devise methods for dealing more effectively with global issues? Will conditions get better or worse? Four possible scenarios of what the world might look like several decades into the future are world government, regionalism, decentralization, and the status quo.

■ WORLD GOVERNMENT

Some scholars argue that a world government, consisting of a powerful central actor with a significant amount of authority, is the method by which we will organize ourselves in the future. The World Trade Organization's increasing degree of consensus on economic issues, and the emergence of a multitude of free trade agreements, are often cited as evidence that some sort of world government is not out of the realm of possibilities. Similarly, for those arguing that the world is moving toward a single global culture (that is, McWorld), a world government might not seem beyond reach.

In contrast to a slow, evolutionary movement toward consensus on issues like economics, it is conceivable that a world government might be created after a catastrophe. An exchange of nuclear attacks or a worldwide economic crisis more destabilizing than the Asian shock (discussed in Chapter 7) might force governments into calling for a central authority that would avoid long, drawn-out negotiations among 190 or more sovereign states.

An obvious problem with the world government scenario, and the reason it will not be realized anytime soon, is the unlikeliness that the countries of the world would voluntarily give up their sovereignty. Furthermore, a world government would face many practical problems, such as who would be responsible for enforcing laws. Would a world government have a powerful military? If so, the fear of tyranny would be realistic. If not, its enforcement capabilities would be questionable.

There are other possibilities in addition to a true world government. A *federation* would establish a relatively weaker world government, similar to the model of the United States, where the federal government shares power with the states. Even weaker would be a *confederation,* in which states would be the dominant actors but would give the world government some jurisdiction. Both federate and confederate systems would give a world government more power than the United Nations currently possesses.

■ REGIONALISM

In the regionalism scenario, countries are organized into groups based on geographic proximity, perhaps following the pattern of current economic groupings like the North American Free Trade Agreement (NAFTA), the European Union (EU), and the Asia Pacific Economic Cooperation (APEC) forum. As with a world government, countries would not completely relinquish their sovereignty, but that sovereignty would likely be significantly reduced. The European Union is the leader in the movement toward economic and political cooperation. Not only has the EU drastically reduced barriers to economic integration and to the movement of people within its borders, it has also adopted a single economic currency and made progress toward a common foreign policy. Although NAFTA and APEC are relatively young in comparison to the EU, their formation represents the current popularity of regional arrangements.

Of course, the regionalism scenario also must deal with the reluctance of countries to relinquish their sovereignty, the fear of concentrating too much power in the hands of a central government, and so on. However, these issues may be easier to resolve in smaller groupings of states than in a world government context.

On the positive side, regionalism would facilitate the coordination of regional policymaking on global issues such as the environment, human rights, and trade. Still, the enhanced ability of countries to coordinate policies within their respective regions would not necessarily translate into cooperation between regions. It could be argued that regionalism would simply transform a world in which *countries* compete into one in which *regions* compete, without solving pressing global problems.

■ DECENTRALIZATION

At the same time that free trade and environmental agreements are being enacted, there is significant evidence of decentralization or disintegration. One example is the strong separatist movement mounted by Canada's French-speaking province of Quebec. The former Soviet Union and the former Yugoslavia are additional examples of disintegration. Although the various separatist movements have differing motives, as Chapter 3 shows, many of them do have in common a desire for self-determination—that is, the desire to break away from the dominant culture and govern themselves. If many of these groups succeed, instead of "one world" or a few regions, hundreds (or perhaps thousands) of new countries could emerge. Each new country would, of course, be smaller and more culturally homogeneous than today's countries, which ideally would alleviate some of the tensions discussed in Chapter 3. However, it also would make achieving international consensus on issues like the environment, human rights, and nuclear proliferation more difficult. These sorts of disintegrative movements have a negative connotation since, in many cases, they involve violence.

There is, however, another type of locally oriented movement, commonly referred to as *civil society,* that has gained momentum in recent years. Civil society comprises nongovernmental, nonprofit groups such as social service providers, foundations, neighborhood watch groups, and religion-based organizations. India's Chipko movement, Kenya's Green Belt movement, the Grameen Bank (all discussed in Chapter 10), and Jubilee 2000 (Chapter 7) are examples of civil society, or *grassroots movements.* Later in this chapter, two more examples of civil society will be discussed. In recent decades, more and more people have turned to civil society, rather than government, to solve their problems.

Reading this book, you may have noticed the many global nongovernmental organizations (NGOs) mentioned. The number of NGOs has increased dramatically, from about 200 in the early 1900s to nearly 5,000 at the end of the twentieth century. Their ranks include Amnesty International, Greenpeace, CARE, the Mennonite Central Committee, and the World Wildlife Federation. Composed of private citizens in more than one country, they focus on such global issues as the environment, poverty, human rights, and peace.

Those frustrated with government's inability to solve global problems insist that centralized governments are not the most effective way to deal with them. Governments, they argue, are simply too far removed from local communities to understand completely the nature of a particular problem and to offer effective solutions. Advocates of civil society are encouraged by the dramatic increase in NGOs. Critics, however, believe local grassroots

efforts will be insufficient to solve global problems like nuclear proliferation, ozone depletion, and global warming. They argue that governments
are the only actors with sufficient resources to effectively confront these
large-scale issues.

■ STATUS QUO

Perhaps the most likely scenario is one in which no dramatic changes occur
over the next several decades. This is not to say that change will be absent,
but that it will be only a gradual continuation of current trends toward globalization in the areas of economic integration, information flow among
countries, the importance of nongovernmental actors (including multinational corporations), and cooperation among countries on environmental
and other issues. Citizens will continue to pledge their allegiance to countries, not economic blocs; states, not groups of private citizens, will remain
the dominant political actors; and short-term domestic interests will prevent
states from surrendering their sovereignty.

    Does this scenario allow an effective response on the part of the global
community to the issues discussed in this book? Critics view the status quo
with suspicion because it has made disappointing progress thus far in such
areas as global warming, peacemaking, and poverty, especially poverty
among children.

■ THE FUTURE: SOURCES OF HOPE AND CONCERN

As many of the chapters in this book point out, there are positive signs in
the world's attempts to deal with the multitude of pressing global issues.
Smallpox appears to have been eliminated. There is a cooperative effort to
deal with ozone depletion. Women have been increasingly successful in
forming effective grassroots movements and making their voices heard. In
the developing world, the United Nations Development Programme reports:

- "A child born today can expect to live eight years longer than one
  born 30 years ago."
- "[The] adult literacy rate [has] increased from an estimated 47% in
  1970 to 73% in 1999."
- "The share of rural families with access to safe water has grown
  more than fivefold." (UNDP 2001: 10)

In addition, progress has been made in health, gender equality, environmental sustainability, and democracy (UNDP 2001: 11).

Despite these successes, there is still a long way to go on a number of issues—many challenges still exist to the creation of a better world. In the remainder of this section, two such challenges are discussed.

Perhaps one of the biggest obstacles we face is created by the North's overconsumption and the South's desire to emulate the lifestyle of the North. As the writers in this book have shown, the North, which constitutes approximately 20 percent of the world's population, is responsible for creating the vast majority of our environmental problems (acid rain, ozone depletion, greenhouse gases, resource depletion, etc.). Meanwhile, the South, which makes up 80 or so percent of the world's population, is trying to imitate the North. If the South industrializes in the same fashion as the North, what will the environmental consequences be? The prospect of hundreds of millions of individuals in the South driving automobiles with leaded gas and no catalytic converters is disturbing from an environmental perspective.

At the heart of this issue is *sustainable development*—the idea that development today should not negatively affect the lives of future generations. Environmentalists have stressed that long-term interests must be given higher priority; for example, forests should not be clear-cut in order to obtain short-term profit, since the indiscriminate clearing of forests is environmentally harmful and will inhibit future development. The North has largely accepted sustainable development in principle but has received criticism from the countries of the South. The latter argue that the North, which is responsible for the vast majority of the world's environmental problems, has no moral right to tell the South that it cannot follow the North's development path. These same critics argue that in many poor countries survival is often at stake, so that environmental concerns must be tied to the issue of current development. Based on this premise, the South insisted that the 1992 Earth Summit focus not only on the environment but also on development, and as a result, the conference was officially called the United Nations Conference on Environment and Development. The South also points out that the North is doing little to adopt a more sustainable lifestyle.

One possible approach to this complex issue is the transfer of advanced, "environmentally friendly" technology from the North to the South. This might allow the South to avoid the adverse affects that accompanied Northern development. The South, however, is not in a position to purchase this technology, and the North has balked at significant transfers. Currently, there is no easy solution to this contentious problem.

A related issue is widespread poverty, another enormous obstacle to a better world. Several chapters in this book highlight the connection between poverty and other issues. Chapter 12 points out that health is directly related to poverty—that the poorer you are, the more likely you are to suffer from disease or malnutrition. Chapters 9 and 10 discuss how the number of children a woman bears will decrease as poverty is alleviated and women gain more control over their lives. The environmental section shows that for those who are desperately poor, issues of immediate survival must take precedence over concerns about the environment. Chapter 2 underscores the vast amounts of money spent on military budgets at the expense of social programs such as health and education.

Central to the issue of poverty is the unequal distribution of wealth, which appears to be getting worse. Chapter 8 highlights this problem. At the same time, as those who live in the North know, poverty is not simply a question of North-South relations: There are many pockets of poverty within the wealthy countries, and evidence suggests that the gap between the rich and poor is increasing within countries as well as between them. And despite the sources of hope mentioned above, the Human Development Report notes, as does Chapter 8, little has been done to alleviate inequality.

Finding a solution to this problem will be difficult. At the domestic level, as Chapter 8 points out, a country must have economic growth before income can be redistributed; but economic growth does not guarantee better income distribution. Chapters 8 and 11 demonstrate that focusing on taxation, education, health care, and other such issues is necessary to foster a more favorable distribution of wealth; however, such an approach typically has little support among those whose wealth would be transferred. The issue becomes even more complex if we confront the *global* distribution of wealth. Within a single country, the wealthy are often taxed at higher rates to support social programs. To attempt to tax the wealthy countries in order to pay for social programs in poorer countries would meet a great deal of opposition, not only from the wealthy in the North, but also from the middle- and lower-income populations. Historically, voluntary aid from North to South has helped somewhat but has been insufficient to seriously address the poverty gap; also, as suggested in Chapter 7, traditional aid may not be the most effective approach to fostering development. As is the case with sustainable development, poverty is a daunting problem.

■ WHAT CAN I DO?

It is important to recognize that the future has not yet been written. The choices that governments, NGOs, and individuals make will have a critical effect on the issues discussed here. Assuming the reader agrees that these issues deserve serious attention, whether on the grounds of self-interest, a sense of patriotism, a religious view, or a sense of humanitarianism (see Chapter 1 for an elaboration of these perspectives), the practical question remains: "What can I do to make a positive difference?"

Common suggestions include: write to your government representatives, vote, buy recycled products, and so on. An increasingly widespread option is to form, join, or support an NGO like those mentioned throughout this book. Below are two brief case studies of NGOs, organized by a small group of people, which have tried to relieve the suffering of many. Both cases deal with the poverty and inequality discussed above. The first case focuses on coffee and the NGO Equal Exchange, and the second case describes the work of Educating for Justice and the manufacturing of tennis shoes.

■ NGO Case Study: Equal Exchange

Throughout the world millions of people are engaged in growing, harvesting, and processing coffee. Most of the coffee growers are individuals who own small plots of land and sell their coffee to middlemen who export the coffee to markets in the North. As a result of the recent drop in coffee prices, the price paid to coffee farmers has dropped to around $0.46 per pound, the lowest price in a century. A more visible result of falling coffee prices is the laying off of Juan Valdez, Colombian coffee's best-known marketing tool. It was recently announced that this coffee icon would be a casualty of advertising cutbacks.

Equal Exchange is one of a few unique companies working to improve the lives of coffee farmers throughout the world. Since 1991 they have helped small coffee farms by creating a market for certified "fair-trade" coffee. Fairly traded products are certified by the TransFair organization and bear its logo. The TransFair designation means, among other things, that farmers will receive a "fair" price for their coffee. Currently, the minimum price paid to farmers for fair trade coffee is $1.26 per pound. This is a significantly better price than they would receive otherwise.

Participating farmers generally have to be organized into a cooperative (a group of many farmers) and make decisions democratically. In addition, cooperatives are expected to pursue sustainable development and reinvest much of the profits into improving living conditions in their community (for example, health care, housing, education). Coffee farms are monitored by outside groups to ensure these practices are followed.

Companies dedicated to fair trade work with a business model that accepts small profits, which enables them to free up more resources for paying the higher coffee prices to farmers.

Apparently enough consumers are making that choice, as Starbucks has recently announced that it will offer "Fair Trade Certified" coffee at its U.S. retail stores. Despite the rapid growth of fair-trade coffee over the past few years, and interest among large companies like Starbucks, fair-trade coffee represents less than 1 percent of U.S. coffee sales. For this growth to continue, consumers will have to make a conscious decision to buy fair-trade coffee. For more information, see the Equal Exchange website at http://www.equalexchange.com/.

■ NGO Case Study: Educating for Justice

The second case also deals with fair wages and individuals in the North seeking solidarity with workers in the South. The story of Educating for Justice (EFJ) began with Jim Keady, an assistant soccer coach at St. John's University seeking his master's degree in theology. While researching a paper on Nike labor practices in 1998, Keady discovered that Nike was violating several tenets of Catholic social teaching. Given St. John's Catholic identity, Keady saw a troubling contradiction in the process of the university's negotiating a multimillion dollar endorsement contract with Nike. After months of trying to seek change through the school's administration, he was essentially forced to resign his position because he could not, in good conscience, follow his boss's order to wear Nike and drop the issue or resign. He charged that because he was forced to be "a walking billboard" for Nike (evidence indicated he was not given the choice of wearing a uniform without the Nike logo), his freedom of speech was being violated.

Although to date his suit has been unsuccessful, Keady has continued with the issue and turned his attention to increasing awareness to what he feels is an exploitative wage paid to Nike workers in Southeast Asia. Nike has responded that no one is forcing these people to work in the factories and that Nike does in fact pay a reasonable wage, since it pays better than many other Indonesian jobs. In reaction to Nike claims, Keady and Leslie Kretzu (cofounder of EFJ) offered to work in a factory manufacturing Nike shoes in order to judge the fairness of Nike's wages. After being turned down, Keady and Kretzu set off for Indonesia to try to live on $1.25 per day, the average wage for those working in Indonesian shoe factories. After spending four weeks living with Indonesian workers, Keady and Kretzu have traveled across the United States telling the story of Indonesian workers and the poor conditions in which they live.

These two activists have targeted Nike because of its high profile and because it is the industry leader. They point out that Nike CEO Phil Knight has a net worth of nearly $6 billion but refuses to pay third world workers a living wage. Other activists criticize Nike for closing its factories in the United States and sending the jobs overseas.

In addition to EFJ's public education program, the organization is also trying an innovative strategy to change Nike's policies. They are raising money in order to buy Nike stock. On the surface this seems like a contradictory policy. However, Nike shareholders are allowed to attend the company's annual shareholders meeting. The hope is that by bringing Indonesian factory workers to attend the shareholders meeting, they can influence other, more traditional shareholders to enact policies more favorable to those producing the shoes.

It appears that the EFJ is making some progress in their crusade to win better working conditions for workers in Asia. Nike has recently admitted some mistakes and has made some minor changes to its policies. To learn more see the following websites: http://www.nikewages.org and http://www.nikebiz.

* * *

Although national and local (and perhaps regional) governments will continue to play important roles, we cannot depend solely on them to solve all of the problems discussed in this book. It is up to each individual to work to create the world he or she prefers. "The most revealing world order statement each of us makes is with his or her life" (Falk, Kim, and Mendlovitz 1982: 14). Dramatic results at the global level can be realized if the world's citizens act. We hope readers will become active and educate others on the issues in this book.

■ DISCUSSION QUESTIONS

1. Which of the future world orders do you think is most likely to emerge? Which do you think is most desirable?
2. Would a strengthened United Nations be desirable?
3. Can you think of another possible world order?
4. What serious challenges, in addition to poverty and the need for sustainable development, do you think confront humanity?
5. What items would you add to the list of things you can do as an individual to make the world a more livable place?
6. What do you think of the efforts of Equal Exchange and Educating for Justice?

■ SUGGESTED READINGS

Brown, Lester R., ed. *State of the World* (annual). New York: W. W. Norton.
Cohen, David, Rosa de la Vega, and Gabrielle Watson (2001) *Advocacy for Social Justice: A Global Action and Reflection Guide.* Bloomfiled, CT: Kumarian Press.
Commission on Global Governance (1995) *Our Global Neighborhood.* New York: Oxford University Press.
Earth Works Group (1989) *50 Simple Things You Can Do to Save the Earth.* Berkeley, CA: Earth Works Press.

Naidoo, Kumi (1999) *Civil Society at the Millennium*. Bloomfiled, CT: Kumarian Press.

Pirages, Dennis C. (1996) *Building Sustainable Societies: A Blueprint for a Post-Industrial World*. Armonk, NY: M. E. Sharpe.

United Nations Development Programme (annual) *Human Development Report*. New York: Oxford University Press.

Acronyms

ABM	Anti-Ballistic Missile Treaty
ACC/SCN	Administrative Committee on Coordination/Sub-Committee on Nutrition
AIDS	Acquired Immunodeficiency Syndrome
AOSIS	Alliance of Small Island States
APEC	Asia Pacific Economic Cooperation
BCE	Before the Common Era
BFW	Bread for the World
CAA	Clear Air Act
CBD	Convention on Biological Diversity
CDC	Centers for Disease Control
CGG	Commission on Global Governance
CNN	Cable News Network
CTBT	Comprehensive Nuclear Test Ban Treaty
CWC	Chemical Weapons Convention
DAC	Development Assistance Committee
DAW	Division for the Advancement of Women
ECSC	European Coal and Steel Community
EEZ	Exclusive Economic Zone
EFJ	Educating for Justice
EPA	Environmental Protection Agency
EPI	Expanded Program of Immunization
EPO	European Patent Office
ERBE	Earth Radiation Budget Experiments
EU	European Union
FAO	Food and Agriculture Organization
FDI	foreign direct investment
FGM	female genital mutilation
FTC	Federal Trade Commission

GATT	General Agreement on Tariffs and Trade
GAVI	Global Alliance for Vaccines and Immunization
GDP	gross domestic product
GNP	gross national product
HCFCs	hydrochlorofluorocarbons
HDI	Human Development Index
HILICs	Highly Indebted Low-Income Countries
HIV	Human Immunodeficiency Virus
ICBM	inter-continental ballistic missiles
ICIDI	Independent Commission on International Development Issues
ICJ	International Court of Justice
ICRC	International Committee of the Red Cross
ICTY	International Criminal Tribunal for the Former Yugoslavia
ICY	International Cooperation Year
IFAD	International Fund for Agricultural Development
IGO	international governmental organization
ILO	International Labour Organization
IMF	International Monetary Fund
INGO	international nongovernmental organization
INSTRAW	International Research and Training Institute for the Advancement of Women
IPA	International Peace Academy
IUCN	World Conservation Union
LDCs	less developed countries
LOS	Law of the Sea Treaty
MDCs	more developed countries
Mercosur	Common Market of the South (Mercado Común del Sur)
MNC	multinational corporation
MONUC	United Nations Organization Mission in the Democratic Republic of Congo
NAFO	North Atlantic Fisheries Organizations
NAFTA	North American Free Trade Agreement
NATO	North Atlantic Treaty Organization
NGO	nongovernmental organization
NICs	newly industrializing countries
NMD	National Missile Defense
NOPEC	non-oil-producing less developed countries
NPT	Non-Proliferation Treaty
NRDC	Natural Resource Defense Council
NTB	nontariff barriers
OAPEC	Organization of Arab Petroleum Exporting Countries
ODA	official development assistance

OPEC	Organization of Petroleum Exporting Countries
PLO	Palestine Liberation Organization
PPP	purchasing power parity
SALT	Strategic Arms Limitations Treaty
SDI	Strategic Defense Initiative
SIDA	Swedish International Development Agency
START	Strategic Arms Reduction Treaty
TRIPs	Trade-Related Aspects of Intellectual Property Rights
UDHR	Universal Declaration of Human Rights
UN	United Nations
UNACC/SCN	UN Administrative Committee Coordination/ Subcommittee on Nutrition
UNAIDS	Joint UN Programme on HIV/AIDS
UNCED	UN Conference on the Environment and Development
UNDESIPA	UN Department for Economic and Social Information and Policy Analysis
UNDP	UN Development Programme
UNDPI	UN Department of Public Information
UNECE	UN Economic Commission for Europe
UNEF	UN Emergency Force
UNEP	UN Environment Programme
UNESCO	UN Economic, Scientific and Cultural Organization
UNFICYP	UN Peacekeeping Force in Cyprus
UNFPA	UN Fund for Population Activities
UNGA	UN General Assembly
UNHCR	UN High Commission for Refugees
UNICEF	UN Children's Fund
UNIFEM	UN Development Fund for Women
UNIFIL	UN Interim Force in Lebanon
UNPD	UN Population Division
USAID	U.S. Agency for International Development
USCC and AN	U.S. Code Congressional and Administrative News
USDJ	U.S. Department of Justice
USDOE	U.S. Department of Energy
USGAO	U.S. General Accounting Office
USSCEPW	U.S. Senate Committee on Environmental and Public Works
USSR	Union of Soviet Socialist Republics
VERs	voluntary export restraints
WCED	World Commission on Environment and Development
WFP	World Food Programme
WHO	World Health Organization
WIC	women, infants, and children

WMD	weapons of mass destruction
WTO	World Trade Organization
WWF	World Wide Fund for Nature or World Wildlife Fund
ZPG	zero population growth

BIBLIOGRAPHY

Abramovitz, Janet N. (2001) "Averting Unnatural Disasters." In Lester R. Brown, Christopher Flavin, and Hilary French, eds., *State of the World 2001*. New York: W. W. Norton.

Agreement on Trade-Related Aspects of Intellectual Property Rights (1994) December 15, 1993, 33 I.L.M. 81.

Ahmed, Samir (1994) "Principles and Precedents in International Law Governing the Sharing of Nile Water." In P. P. Howell and J. A. Allan, eds., *The Nile: Sharing a Scarce Resource*. New York: Cambridge University Press.

Altman, D. G., et al. (1996) "Tobacco Promotion and Susceptibility to Tobacco Use Among Adolescents," *American Journal of Public Health* 86, no. 11.

Amler, R. W., and H. B. Dull (1987) *Closing the Gap*. Oxford: Oxford University Press.

Anand, Anita (1983) "Saving Trees, Saving Lives: Third World Women and the Issue of Survival." In Leonia Caldecott and Stephanie Leland, eds., *Reclaim the Earth: Women Speak Out for Life on Earth*. London: Women's Press.

Aspin, Les (1994) *Secretary of Defense Annual Report to the Congress and President*. Washington, DC: Government Printing Office.

Baker, Christopher (1999) *Costa Rica Handbook*. Third edition. New York: Moon Publications. Available online at www.photo.net/cr/moon/conservation.html (October 6, 2001).

Barber, Benjamin R. (1992) "Jihad vs. McWorld," *Atlantic Monthly,* March.

———— (1996) *Jihad vs. McWorld*. New York: Ballantine Books.

Barr, Cameron W. (2001) "Gulf Legacy: US Quietly Guards Oil." *Christian Science Monitor* 93, no. 64 (February 27): 1.

Bates, A. K. (1990) *Climate in Crisis*. Summertown, TN: Book Publishing Company.

Bell, Daniel A. (1996) "The East Asian Challenge to Human Rights: Reflections on an East West Dialogue," *Human Rights Quarterly* 18, no. 3.

Berkes, F., D. Feeny, B. J. McCay, and J. M. Acheson (1989) "The Benefits of the Commons," *Nature* 340 (July 13): 91–93.

BFW (Bread for the World) Institute (1994) *Hunger 1995: Causes of Hunger.* Washington, DC: BFW Institute.

———— (1995) "At the Crossroads: The Future of Foreign Aid." Occasional Paper, No. 4. Washington, DC: Bread for the World Institute.

——— (1997) *Hunger in a Global Economy: Hunger 1998*. Washington, DC: BFW Institute.

Birdsall, Nancy, Thomas Pinckney, and Richard Sabot (1996) "Why Low Inequality Spurs Growth: Savings and Investment by the Poor." Inter-American Development Bank Working Paper Series, No. 327. Washington, DC: IDB.

Birdsall, Nancy, David Ross, and Richard Sabot (1995) "Inequality and Growth Reconsidered: Lessons from East Asia," *World Bank Economic Review* 9, no. 3.

Blecker, Robert A. (1999) *Taming Global Finance*. Washington, DC: Economic Policy Institute.

Bojtar, Endre (1988) "Eastern or Central Europe?" *Cross Currents* 7.

Boserup, Ester (1970) *Women's Role in Economic Development*. New York: St. Martin's Press.

——— (1981) *Population and Technological Change: A Study of Long Term Trends*. Chicago: University of Chicago Press.

Boston Globe (1996) Special Report, "Armed for Profit: The Selling of U.S. Weapons," February 11.

Bouhdiba, Abdelwahab (1982) *Exploitation of Child Labour*. New York: United Nations.

Boulding, Elise (1992) *The Underside of History: A View of Women Through Time*. Revised edition. Newbury Park, CA: Sage Publications.

Boutros-Ghali, Boutros (1992) *An Agenda for Peace: Preventive Diplomacy, Peacemaking and Peace-Keeping*. New York: United Nations.

——— (1995) *An Agenda for Peace*. Second edition. New York: United Nations.

Braudel, F. (1981) *The Structures of Everyday Life: Civilization and Capitalism— 15th–18th Century*. Vol. 1. New York: Harper & Row.

Breuilly, John (1993) *Nationalism and the State*. Second edition. Chicago: University of Chicago Press.

Bright, Chris (2000) "Anticipating Environmental Surprise." In Lester R. Brown, Christopher Flavin, and Hilary French, eds., *State of the World 2000*. New York: W. W. Norton.

Brown, Lester R., ed. (1993) "A New Era Unfolds." In Lester R. Brown, ed., *State of the World*. New York: W. W. Norton.

———, ed. (1996) "The Acceleration of History." In Lester R. Brown, ed., *State of the World*. New York: W. W. Norton.

Bundy, McGeorge, William J. Crowe, Jr., and Sidney Drell (1993) *Reducing Nuclear Danger: The Road Away from the Brink*. New York: Council on Foreign Relations Press.

Burstyn, Linda (1995) "Female Circumcision Comes to America," *Atlantic Monthly*, October.

Byrd, Veronica (1994) "The Avon Lady of the Amazon," *Business Week*, October 24.

Caldwell, John C. (1982) *Theory of Fertility Decline*. New York: Academic Press.

Caldwell, Lynton K. (1990) *International Environmental Policy: Emergence and Dimensions*. Second edition. Durham, NC: Duke University Press.

Carson, Rachel (1962) *Silent Spring*. New York: Fawcett Crest.

Castles, Stephen, and Mark J. Miller (1993) *The Age of Migration: International Population in the Modern World*. New York: Guilford Press.

——— (1998) *The Age of Migration: International Population in the Modern World*, Second edition. New York: Guilford Press.

Cavanaugh, John, et al., eds. (1992) *Trading Freedom: How Free Trade Affects Our Lives, Work, and Environment*. San Francisco: Institute for Food and Development Policy.

Center for Defense Information (2001) "World Military Expenditures: US vs. the World." Available online at www.cdi.org/issues/wme (July 24).

CGG (Commission on Global Governance) (1995) *Our Global Neighborhood*. New York: Oxford University Press.

CIOSC (China, Information Office of the State Council) (2001) *US Human Rights Record in 2000*, Beijing. Available online at http://www.chinadaily.com.cn/highlights/paper/us2000.html (February 27).

Claude, Richard Pierre, and Burns H. Weston, eds. (1992) *Human Rights in the World Community*. Philadelphia: University of Pennsylvania Press.

Cohen, William S. (2001) *Secretary of Defense Annual Report to the President and Congress*. Washington, DC: Government Printing Office.

Columbus Dispatch (1993) "U.N. Group Begins War on Female Circumcision," May 13.

Commoner, Barry (1992) *Making Peace with the Planet*. New York: New Press.

Crossette, Barbara (2001) "U.N. Effort to Cut Arms Traffic Meets a U.S. Rebuff," *New York Times,* July 10, p. 1.

Davis, Zachary S. (1991) *Non-Proliferation Regimes: A Comparative Analysis of Policies to Control the Spread of Nuclear, Chemical, and Biological Weapons and Missiles*. Washington, DC: Congressional Research Service.

DeMont, John (1995a) "Gunboat Diplomacy," *Maclean's,* March 20.

——— (1995b) "Reeling in a Deal," *Maclean's,* April 24.

Desai, Narayan (1972) *Toward a Nonviolent Revolution*. Varanasi, India: Sarva Seva Sangh Prakashan.

Dobson, Wendy, and Gary Clyde Hufbauer (2001) *World Capital Markets*. Washington, DC: Institute for International Economics.

Dodge, Robert (1994) "Grappling with GATT," *Dallas Morning News,* August 8.

Donnelly, Jack (1993) *International Human Rights*. Boulder, CO: Westview Press.

Drake, Christine (2000) "Water Resource Conflicts in the Middle East." *World & I* 15, no. 9 (September): 298+.

Dregne, H. E., ed. (1991) *Degradation and Restoration of Arid Lands*. Lubbock: Texas Tech University, International Center for Arid and Semiarid Land Studies.

Drêze, Jean, and Amartya Sen (1989) *Hunger and Public Action*. Oxford: Clarendon Press.

Drinan, Robert F. (1987) *Cry of the Oppressed*. San Francisco: Harper & Row.

Dunn, Seth (2001a) "Decarbonizing the Energy Economy." In Lester R. Brown, Christopher Flavin, and Hilary French, eds., *State of the World 2000*. New York: W. W. Norton.

——— (2001b) "Atmospheric Trends." In *The Worldwatch Institute: Vital Signs 2001*. New York: W. W. Norton.

Easterly, William (1999) "The Lost Decades: Explaining Developing Countries' Stagnation 1980–1998." Washington, DC: World Bank Policy Research Working Paper.

Ehrlich, Paul, and Anne Ehrlich (1992) *The Population Explosion*. New York: Doubleday.

Elliott, Jennifer A. (1994) *An Introduction to Sustainable Development: The Developing World*. London: Routledge.

Fairclough, Gordon (2001) "Should Trade Have a No-Smoking Section?" *Wall Street Journal,* July 23.

Faison, Seath (1997) "China Turns the Tables, Faulting U.S. on Rights," *New York Times,* March 5.

Falk, Richard, Samuel S. Kim, and Saul H. Mendlovitz (1982) *Toward a Just World Order.* Vol. 1. Boulder, CO: Westview Press.

Fallows, James (1993) "How the World Works," *Atlantic Monthly,* December.

—— (1996) *The Sixth World Food Survey.* Rome: FAO.

—— (2000) *The State of the World Fisheries and Aquaculture 2000.* Available online at www.fao.org/docrep/003/x8002e00.htm (October 14, 2001).

FAO (Food and Agriculture Organization) (1996) *Occasional World Food Survey.* Rome: FAO.

—— (2000) "The State of World Fisheries and Agriculture." Accessed online October 14, 2001, at www.fao.org/docrep/002/x8002e00.htm.

—— (2001) *The State of Food Insecurity in the World.* Rome: FAO.

Farer, Tom J. (1992) "The United Nations and Human Rights: More Than a Whimper, Less Than a Roar." In Richard Pierre Claude and Burns H. Weston, eds., *Human Rights in the World Community.* Philadelphia: University of Pennsylvania Press.

Felice, William F. (1996) *Taking Suffering Seriously.* Albany: State University of New York Press.

Filkins, Dexter (2001) "In Fallen Taliban City, a Busy, Busy Barber." *New York Times,* November 13.

Fiske, Edward B. (1993) *Basic Education: Building Block for Global Development.* Washington, DC: Academy for Educational Development.

Flavin, Christopher (1991) "Conquering U.S. Oil Dependence," *Worldwatch* 4, no. 1.

—— (1996) "Facing Up to the Risks of Climate Change." In Lester R. Brown, ed., *State of the World.* New York: W. W. Norton.

Flavin, Christopher, and Seth Dunn (1997) "Rising Sun, Gathering Winds: Policies to Stabilize the Climate and Stengthen Economies." In *Worldwatch Paper* No. 138.

Flavin, Christopher, and Nicholas Lenssen (1991) "Designing a Sustainable Energy System." In Lester R. Brown, ed., *State of the World.* New York: W. W. Norton.

Forsberg, Randall, William Driscoll, Gregory Webb, and Jonathan Dean (1995) *Nonproliferation Primer: Preventing the Spread of Nuclear, Chemical, and Biological Weapons.* Cambridge, MA: MIT Press.

Forsyth, Randall W. (1996) "The End of Communism, the End of History and Now, the End of Business Cycles?" *Barron's,* November 18.

Forsythe, David P. (1991) *The Internationalization of Human Rights.* Lexington, MA: Lexington Books.

Freedman, Lawrence, and Efraim Karsh (1993) *The Gulf Conflict, 1990–1991: Diplomacy and War in the New World Order.* Princeton, NJ: Princeton University Press.

French, Hilary, and Lisa Mastny (2001) "Controlling International Environmental Crime." In Lester R. Brown, Christopher Flavin, and Hilary French, eds., *State of the World 2000.* New York: W. W. Norton.

Friedman, Thomas L. (1997) "Rethinking China, Part I." *New York Times,* March 3.

Galtung, Johan (1969) "Violence, Peace, and Peace Research," *Journal of Peace Research* 6, no. 6.

Garrett, Laurie (1994) *The Coming Plague.* New York: Farrar, Straus & Giroux.

Ghazi, Polly, Frank Smith, and Claire Trevena (1995) "Our Plundered Seas," *World Press Review* 42, no. 6.

Gleick, Peter H. (1994) "Water, War and Peace in the Middle East," *Environment* 36, no. 3.

Gore, Al (1992) *Earth in the Balance: Ecology and the Human Spirit.* Boston: Houghton Mifflin.

Graham, Edward M. (1996) *Global Corporations and National Governments*. Washington, DC: Institute for International Economics.

Gribbin, J. (1988) *The Hole in the Sky: Man's Threat to the Ozone Layer*. New York: Bantam Books.

Gudeman, Stephen (1996) "Sketches, Qualms, and Other Thoughts on Intellectual Property Rights." In Stephen B. Brush and Doreen Stabinsky, eds., *Valuing Local Knowledge*. Washington, DC: Island Press, pp. 102–121.

Harbottle, Michael (1971) *The Blue Berets*. London: Leo Cooper.

Hardin, Garrett (1968) "The Tragedy of the Commons," *Science* 162 (December 13): 1243–1248.

Harper, Charles L. (1995) *Environment and Society: Human Perspectives on Environmental Issues*. Upper Saddle River, NJ: Prentice Hall.

Hartung, William D. (1995) *And Weapons for All*. New York: HarperCollins.

Hayner, Priscilla B. (2001) *Unspeakable Truths: Confronting State Terror and Atrocity*. New York and London: Routledge.

Heilbroner, Robert L. (1985) *The Nature and Logic of Capitalism*. London: Norton.

Hersh, Seymour M. (1994) "The Wild East," *Atlantic Monthly*, July.

Higgins, Rosalyn (1996) *United Nations Peacekeeping 1946–1967*. Vol. 1, *The Middle East*. London: Oxford University Press.

Hillel, Daniel (1994) *Rivers of Eden: The Struggle for Water and the Quest for Peace in the Middle East*. New York: Oxford University Press.

Hindustan Times (1998) "Neem Goes Global!" *Hindustan Times Sunday Magazine*, October 18. Available at the *Hindustan Times* website, http://www.hindustantimes.com/nonfram/181098/SUN06.htm.

Hobbelink, Henk (1991) *Biotechnology and the Future of World Agriculture*. London: Zed Books.

Holdren, John P., and Paul R. Ehrlich (1974) "Human Population and the Global Environment," *American Scientist* 62 (May).

Holm, Hans-Henrik, and Georg Sørensen, eds. (1995) *Whose World Order? Uneven Globalization and the End of the Cold War*. Boulder, CO: Westview Press.

Hottelet, Richard C. (1996) "Syria Tries to Shore Up Weak Position," *Christian Science Monitor* 88, no. 135 (June 7): 18.

Hoy, Paula (1998) *Players and Issues in International Aid*. Bloomfield, CT: Kumarian Press.

Human Rights Watch (1988) "The Persecution of Human Rights Monitors: December 1987 to December 1988, A Worldwide Survey." New York: Human Rights Watch.

Huntington, Samuel P. (1996) "The West: Unique, Not Universal," *Foreign Affairs* 75, no. 6.

ICDSI (Independent Commission on Disarmament and Security Issues) (1982) *Common Security*. New York: Simon & Schuster.

ICIDI (Independent Commission on International Development Issues) (1980) *North-South: A Programme for Survival*. Cambridge, MA: MIT Press.

ICJ (International Court of Justice) (1998) "Case Concerning Fisheries Jurisdiction (Spain v. Canada)," Summary of the Judgment. December 4, 1998. Available online at www.icj-cij.org/icjwww/idocket/iec/iec_summaries/iecsummary/19981204.htm/ (October 14, 2001).

ICRC (International Committee of the Red Cross) (1993) "A Time for Decision," *International Review of the Red Cross* (November–December).

IFOAM (International Federation of Organic Agriculture Movements) (1997) "Good News in Two IFOAM Actions on Genetic Engineering." Press release dated December 8, available online at http://ecoweb.dk/ifoam/gmo/pr971208.htm.

IISS (International Institute for Strategic Studies) (2000) *The Military Balance, 1999–2000*. London: IISS.

ILO (International Labour Organization) (1976) *Wages and Working Conditions in Multinational Enterprises*. Geneva: ILO.

——— (1993) *World Labour Report*. Geneva: ILO.

——— (1996) *World Employment: 1996/97*. Geneva: ILO.

IMF (International Monetary Fund) (1991) *International Capital Markets, Developments and Prospects*. Washington, DC: IMF.

IPA (International Peace Academy) (1984) *Peacekeeper's Handbook*. New York: Pergamon Press.

IUCN (World Conservation Union), the United Nations Environment Programme (UNEP), and the World Wide Fund for Nature (WWF) (1991) *Caring for the Earth*. Gland, Switzerland: IUCN, UNEP, and WWF.

Jenkins, Christopher, et al. (1997) "Tobacco Use in Vietnam," *Journal of the American Medical Association* 277, no. 21.

Jones, Jeffrey R. (1992) "Environmental Issues and Policies in Costa Rica: Control of Deforestation," *Policy Studies Journal* 20, no. 4.

Jones, Rodney W., Mark G. McDonough, Toby F. Dalton, and Gregory D. Koblentz (1998) *Tracking Nuclear Proliferation: A Guide in Maps and Charts*. Washington, DC: Carnegie Endowment for International Peace.

Juma, Calestous (1989) *The Gene Hunters: Biotechnology and the Scramble for Seeds*. Princeton, NJ: Princeton University Press.

Kahn, Jeremy (2000) "The World's Largest Corporations," *Fortune Magazine* 142, July 24.

Karp, Aaron (1994) "The Arms Trade Revolution: The Major Impact of Small Arms," *Washington Quarterly* 17 (autumn).

Kegley, Charles W., and Eugene R. Wittkopf (1997) *World Politics: Trend and Transformation*. New York: St. Martin's Press.

Kent, George (1995) *Children in the International Political Economy*. New York: St. Martin's Press.

Kerr, Richard A. (1989) "Greenhouse Skeptic Out in the Cold," *Science*, December.

——— (2000) "Can the Kyoto Climate Treaty Be Saved from Itself?" *Science*, November 3.

Khor, Martin (1997) "Rio+5 and the Global Ecological Crisis." Speech given June 27. Available online at http://www.globalpolicy.org/socecon/envronmt/khor.htm.

——— (1998) "A Worldwide Fight Against Biopiracy and Patents on Life." Available online at the Third World Network website, http://www.twnside.org.sg/souths/twn/title/pat-ch.htm (November 16).

Klare, Michael (1999) "The Kalashnikov Age," *Bulletin of the Atomic Scientists* (January–February).

Klinger, Janeen (1994) "Debt-for-Nature Swaps and the Limits to International Cooperation on Behalf of the Environment," *Environmental Politics* (summer).

Kohn, Hans (1965) *Nationalism: Its Meaning and History*. Revised edition. New York: Van Nostrand Reinhold.

Korten, David C. (1996) *When Corporations Rule the World*. West Hartford, CT: Kumarian Press.

——— (2001) *When Corporations Rule the World*. Second edition. Bloomfield, CT: Kumarian Press.

Lamar, B. (1991) "Life Under the Ozone Hole: In Chile, the Mystery of the Bug-Eyed Bunnies," *Newsweek*, December 9.

Landres, Shawn (1996) "For God and Country: The Importance of Religion in the Study of Nationalities," *ASNews: The Newsletter of the Association for the Study of Nationalities* 2, no. 3 (fall).

Langley, Winston E. (1996) *Encyclopedia of Human Rights Issues Since 1945.* Westport, CT: Greenwood Press.

Laurance, Edward J. (1992) *The International Trade in Arms,* New York: Lexington Books.

Leggett, J. (1990) "The Nature of the Greenhouse Threat." In J. Leggett, ed., *Global Warming: The Greenpeace Report.* New York: Oxford University Press.

Liebich, André, Daniel Warner, and Jasna Dragovic, eds. (1995) *Citizenship East and West.* London: Kegan Paul International.

Lindzen, R. (1993) "Absence of Scientific Basis," *Research and Exploration* (spring).

Lok Swaasthya Parampara Samvardhan Samithi (n.d.) "A Case History of Biopiracy." In *Neem: A User's Manual.* Available online at the Health Education Library for People website, http://www.healthlibrary.com/reading/neem/chap10.htm.

Lutz, Wolfgang (1994) "World Population Trends: Global and Regional Interactions Between Population and Environment." In Arizpe M. Lourdes, Priscilla Stone, and David C. Major, eds., *Population and Environment: Rethinking the Debate.* Boulder, CO: Westview Press.

MacKenzie, James J. (2001) "Facing the United States' Oil Supply Problems: Would Opening Up the Arctic National Wildlife Refuge (ANWR) Coastal Plain Really Make a Difference?" Available online at www.wri.org/climate/anwr.html (October 21).

Mahony, Liam, and Luis Enrique Eguren (1997) *Unarmed Bodyguards.* West Hartford, CT: Kumarian Press.

Mahony, Rhona (1992) "Debt-for-Nature Swaps: Who Really Benefits?" *Ecologist* 22, no. 3.

Malthus, T. R. (1826) *An Essay on the Principle of Population.* Sixth edition. London: Reeves & Turner.

Matthews, David Ralph (1996) "Mere Anarchy? Canada's 'Turbot War' as the Moral Regulation of Nature," *Canadian Journal of Sociology* 21, no. 4 (fall): 505–522.

McCarthy, Sheryl (1996) "Fleeing Mutilation, Fighting for Asylum," *Ms.,* July–August.

McCormick, John (1989) *Reclaiming Paradise: The Global Environmental Movement.* Bloomington: Indiana University Press.

McGwire, Michael (1994) "Is There a Future for Nuclear Weapons?" *International Affairs* 70, no. 2.

McKibben, B. (1989) *The End of Nature.* New York: Anchor Books.

McKinney, M. L., and R. M. Schoch (1996) *Environmental Science: Systems and Solutions.* Minneapolis/St.Paul: West Publishing.

McNaugher, Thomas L. (1990) "Ballistic Missiles and Chemical Weapons," *International Security* 15, no. 2.

Meadows, D. H., D. L. Meadows, and J. Rander (1992) *Beyond the Limits: Confronting Global Collapse, Envisioning a Sustainable Future.* Mills, VT: Chelsea Green Publishers.

Michaels, P. (1992) *Sound and Fury: Science and Politics of Global Warming.* Washington, DC: Cato Institute.

Michaels, Patricia A. (n.d.) "Debt for Nature Swaps." Available online at http://environment.about.com/library/weekly/aa091700.htm (September 26, 2001).

Michalak, Stanley (2001) *A Primer in Power Politics.* Wilmington, DE: Scholarly Resources, Inc.

Milanovic, Branko (2001) World Bank Working Paper No. 2244. *True World Income Distribution, 1988 and 1993: First Calculations, Based on Household Surveys Alone.* Washington, DC: World Bank.

Miller, Marian A. L. (1995) *The Third World in Global Environmental Politics.* Boulder: Lynne Rienner.

———— (2001) "Tragedy for the Commons: The Enclosure and Commodification of Knowledge." In Dimitris Stevis and Valerie J. Assetto, eds., *The International Political Economy of the Environment.* Boulder: Lynne Rienner, pp. 111–134.

Moon, Bruce E. (1996) *Dilemmas of International Trade.* Boulder, CO: Westview Press.

———— (1998) "Exports, Outward-Oriented Development, and Economic Growth," *Political Research Quarterly* (March).

———— (2000) *Dilemmas of International Trade.* Second edition. Boulder, CO: Westview Press.

Moser-Puangsuwan, Yeshua, and Thomas Weber (2000*) Nonviolent Intervention Across Borders.* Honolulu: University of Hawaii Press.

Mowlana, Hamid (1995) "The Communications Paradox," *Bulletin of Atomic Scientists* 51, no. 4.

Murray, C. J. L., and A. D. Lopez, eds. (1996) *Global Burden of Disease.* Cambridge, MA: Harvard University Press.

Nagy, Thomas J. (2001) "The Secret Behind the Sanctions." *Progressive* 65, no. 9 (September): 22+.

National Research Council (1986) *Population Growth and Economic Development: Policy Questions.* Washington, DC: National Academy Press.

Nature Conservancy (2001) "Landmark Deal to Protect Rainforests in Belize." Available online at http://nature.org/wherewework/centralamerica/belize/press/press351.html (September 26).

Newhouse, John (2001) "The Missile Defense Debate," *Foreign Affairs* 80, no. 4 (July–August).

New York Times (1996) "Third World Debt Crisis," June 28.

———— (2000) "The United Nations, China Sign Human-Rights Pact." November 20.

———— (2001) "Milosevic Charged with Bosnia Genocide," July 23.

Nincic, Miroslav (1982) *The Arms Race: The Political Economy of Military Growth.* New York: Praeger.

"Nor Any Drop to Drink" (2001) *Lancet* 358, no. 9287 (September 29): 1025.

NRDC (Natural Resources Defense Council) (2001) "Reducing U.S. Oil Dependence." Available online at www.nrdc.org/air/energy/fensec.asp (October 21).

Ostrom, Elinor (1990) *Governing the Commons: The Evolution of Institutions for Collective Action.* Cambridge, UK: Cambridge University Press.

Patel, Surendra J. (1996) "Can the Intellectual Property Rights System Serve the Interests of Indigenous Knowledge?" In Stephen B. Brush and Doreen Stabinsky, eds., *Valuing Knowledge.* Washington, DC: Island Press, pp. 305–322.

Payne, Stanley (1995) *A History of Fascism 1914–1945.* Madison: University of Wisconsin.

"Peace Walks on Water" (2000) *Economist* 356, no. 8190 (September 30): 45.

Peterson, Scott (2000) "Turkey's Plan for Mideast Peace," *Christian Science Monitor* 92, no. 102 (April 18): 1.

Pickering, Kevin T., and Lewis A. Owen (1994) *An Introduction to Global Environmental Issues.* London: Routledge.

Polanyi, Karl (1944) *The Great Transformation.* New York: Farrar & Reinhart.

Pomfret, Richard (1988) *Unequal Trade.* Oxford: Basil Blackwell.

Postel, Sandra (1993) "The Politics of Water," *Worldwatch* 6, no. 4.

―――― (1994) "Carrying Capacity: Earth's Bottom Line." In Lester R. Brown, ed., *State of the World.* New York: W. W. Norton.

Powelson, John P. (1977) "The Oil Price Increase: Impacts on Industrialized and Less-Developed Countries," *Journal of Energy and Development* (autumn).

Prashad, Vijay (1994) "Contract Labor: The Latest Stage of Illiberal Capitalism." *Monthly Review* 46, no. 5: 19–26.

Public Policy Center (2000) "Beef Imports from Brazil." Available online at http://hill.beef.org/files/FSPP/brazil.htm (April).

Raj, R. Dev. (1998) "Trade-India: Sifting 'Basmati' Grain from Patents Chaff." *Interpress Service World News.* Report dated March 17, available online at http://www.oneworld.org/ips2/mar98/basmati.html.

Ramsbottom, Oliver, and Tom Woodhouse (1999*) Encyclopedia of International Peacekeeping Operations.* Santa Barbara, CA: ABC-CLIO, Inc.

Rathjens, George (1995) "Rethinking Nuclear Proliferation," *Washington Quarterly* 18 (winter).

Ravallion, Martin (2001) World Bank Policy Research Working Paper No. 2558. *Growth, Inequality, and Poverty: Looking Beyond Averages.* Washington DC: World Bank.

Ravallion, Martin, and Shaohua Chen (1997) "What Can New Survey Data Tell Us About Recent Changes in Distribution and Poverty?" *World Bank Economic Review.* Washington, DC: World Bank.

Ray, D. L., and L. Guzzo (1992) *Trashing the Planet.* New York: Harper Perennial.

Redclift, Michael (1987) *Sustainable Development: Exploring the Contradictions.* New York: Methuen.

Redfern, Paul (1995) "Africa: Left Out?" *East Africa* (October 30–November 5). Quoted in *World Press Review,* June 6, 1996.

Renan, Ernest (1996) "What Is a Nation?" In Geoff Eley and Ronald Grigor, eds., *Becoming National: A Reader.* New York: Oxford University Press.

Ricardo, David (1981) *Works and Correspondence of David Ricardo: Principles of Political Economy and Taxation.* London: Cambridge University Press.

Roberts, Brad (1995) *Weapons Proliferation in the 1990s.* Cambridge, MA: MIT Press.

Robinson, Geoffrey (2000) *Crimes Against Humanity: The Struggle for Global Justice.* New York: W. W. Norton.

Roth, Kenneth (2001) "Milosevic's Indictment Sets Much-Needed Precedent," *Miami Herald,* July 13.

Rourke, John T. (1995) *International Politics on the World Stage.* Fifth edition. Guilford, CT: Dushkin Publishing Group.

―――― (1997) *International Politics on the World Stage.* Sixth edition. Guilford, CT: Dushkin/McGraw Hill.

Russell, Dick (1995) "High-Seas Fishing: Lawless No Longer," *Amicus Journal* 17, no. 3.

Ryan, John C. (1992) "Conserving Biological Diversity." In Lester R. Brown, ed., *State of the World 1992.* New York: W. W. Norton.

Sagan, Scott D. (1986) "1914 Revisited: Allies, Offense, and Instability," *International Security* 11, no. 2.

Sarkar, Amin U., and Karen L. Ebbs (1992) "A Possible Solution to Tropical Troubles?" *Futures* 24, no. 7.

Scheffer, David J. (1996) "International Judicial Intervention," *Foreign Policy* 102 (spring).

Scheumann, Waltina (1998) "Conflicts on the Euphrates: An Analysis of Water and Non-Water Issues." In W. Scheumann and M. Schiffler, eds., *Water in the Middle East: Potential for Conflicts and Prospects for Cooperation*. Berlin: Springer.

Schlesinger, James (1993) "The Impact of Nuclear Weapons on History," *Washington Quarterly* 16 (autumn).

Schumacher, E. F. (1993) *Small Is Beautiful: Economics as If People Mattered*. San Francisco: Harper & Row.

Seis, M. (1996) "An Eco-Critical Criminological Analysis of the 1990 Clean Air Act." Ph.D. diss., Indiana University of Pennsylvania.

"Sharing Mesopotamia's Water" (1999) *Economist* 353, no. 8145 (November 13): 43+.

Shilts, Randy (1987) *And the Band Played On*. New York: St. Martin's Press.

Shiva, Vandana (n.d.) "The Neem Tree: A Case History of Biopiracy." Available online at the Third World Network website: http://www.twnside.org.sg/souths/twn/title/pir-ch.htm.

———— (1997) *Biopiracy: The Plunder of Nature and Knowledge*. Boston: South End Press.

Shiva, Vandana, and Radha Holla-Bhar (1996) "Piracy by Patent: The Case of the Neem Tree." In Jerry Mander and Edward Goldsmith, eds., *The Case Against the Global Economy*. San Francisco: Sierra Club Books, pp. 146–159.

Simon, Julian L. (1990) *Population Matters: People, Resources, Environment, and Immigration*. New Brunswick, NJ: Transaction Publishers.

SIPRI (Stockholm International Peace Research Institute) (2001) *SIPRI Yearbook 2001: Armaments, Disarmament, and International Security*. London: Oxford University Press.

Sivard, Ruth Leger (1991) *World Military and Social Expenditures 1991*. Washington, DC: World Priorities.

Small, Melvin, and J. David Singer (1982) *Resort to Arms: International and Civil Wars, 1816–1980*. Beverly Hills, CA: Sage Publications.

Smith, Adam (1910) *An Inquiry into the Nature and Causes of the Wealth of Nations*. London: J. M. Dutton.

Steinberg, Gerald M. (1994) "U.S. Non-Proliferation Policy: Global Regimes and Regional Realities," *Contemporary Security Policy* 15, no. 3.

Stevenson, Glenn G. (1991) *Common Property Economics: A General Theory and Land Use Applications*. Cambridge: Cambridge University Press.

Sugar, Peter F., and Ivo John Lederer, eds. (1994) *Nationalism in Eastern Europe*. Seattle: University of Washington Press.

Switzer, Jacqueline Vaughn (1994) *Environmental Politics: Domestic and Global Dimensions*. New York: St. Martin's Press.

Teich, Mikuláš, and Roy Porter, eds. (1993) *The National Question in Europe in Historical Context*. Cambridge: Cambridge University Press.

"Thailand Threatened Over Intellectual Property Law" (1997) *The Ecologist* 27, no. 3: C3.

Toffler, Alvin, and Heidi Toffler (1991) "Economic Times Zones: Fast Versus Slow," *New Perspectives Quarterly* 8, no. 4.

Tucker, Jonathan B. (2000) "Chemical and Biological Terrorism: How Real a Threat?" *Current History*, April.

Tumulty, Brian (1994) "U.S. Industry Confronts Cost of Implementing GATT," Gannett News Service, July 18.

"Turkey's Latest Controversial Dam" (2000) *Economist* 355, no. 8168 (April 29): 51.

Tyler, Patrick E. (1997) "China and Red Cross Agree to New Talks on Prison Visits," *New York Times*, February 29.

UN (United Nations) (1972) *Report of the United Nations Conference on the Human Environment Held at Stockholm, 5–16 June 1972*. List of Participants, UN Doc. A/Conf. 48/Inf. 5.

——— (1973) "Report of the Secretary-General on the Implementation of the Security Council Resolution 340." UN document S/11052/Rev., October 27.

——— (1976) "World Plan of Action, Report of the World Conference of the International Women's Year, Mexico City, June 19–July 1, 1975." F/Conf. 66/34. New York: United Nations.

——— (1988) *Human Rights: Questions and Answers*. New York: United Nations.

——— (1992) *Agenda 21: Programme of Action for Sustainable Development*. New York: United Nations Department of Public Information.

——— (1998) "United Nations Press Briefing on Kyoto Protocol," March 16.

UNACC/SCN (United Nations Administrative Committee on Coordination/Subcommittee on Nutrition) (1997) "Update on the Nutrition Situation 1996: Summary of Results for the Third Report on the World Nutrition Situation." Geneva: ACC/SCN.

UNAIDS (2001) *AIDS Epidemic Update—December 2000*. Available online at http://www.unaids.org/epidemic_update/report_dec00/index_dec.html#full.

UNDESIPA (United Nations Department for Economic and Social Information and Policy Analysis, Population Division) (1994) "The Sex and Age Distribution of the World Populations." New York: United Nations.

——— (1995) "World Urbanization Prospects: The 1994 Revision." New York: United Nations.

UNDP (United Nations Development Programme) (1994) *Human Development Report*. New York: Oxford University Press.

——— (1996) *Human Development Report*. New York: Oxford University Press.

——— (1997) *Human Development Report*. New York: Oxford University Press.

——— (1998) "Debt-for-Environment Swaps for National Desertification Funds." Available online at www.undp.org/seed/unso/pub-htm/swap-eng1.htm (September 26, 2001).

——— (2000) *Global Population Policy: Database, 1999*. New York: United Nations.

——— (2001) *Human Development Report*. New York: Oxford University Press.

UNDPI (United Nations Department of Public Information) (1990) *United Nations Peace-Keeping*. DPI/1048, May. New York: UNDPI.

——— (1996) *The Blue Helmets: A Review of United Nations Peace-Keeping*. Third edition. DPI 1800, Sales No.: E.96.I.14. New York: UNDPI.

——— (1998) *Peacekeeping: 50 Years 1948–1998*. Pamphlet DPI/2004, 88 pages. New York: UNDPI.

——— (2000) *Report of the Panel on UN Peace Operations* (the "Brahimi Report"). UN Resolutions A/55/305-S/2000/809 (August 21).

UNEP (United Nations Environment Programme) (1996) *Global Biodiversity Assessment*. Nairobi: UNEP.

UNGA (United National General Assembly) (1947) Resolution 109 (II), October 21.

——— (1948) Resolution 186 (ES-1), May 14.

——— (2001) *We the Children: End-Decade Review of the Follow-Up to the World Summit for Children: Report of the Secretary General*. A/S-27/3. New York: United Nations.

UNHCR (United Nations High Commissioner for Refugees) (1995) *The State of the World's Refugees 1995*. Oxford: Oxford University Press.

——— (2001) Population Data Unit. *Refugees by Numbers, 2000 Edition*.

UNICEF (United Nations Children's Fund) (various years) *The State of the World's Children*. New York: Oxford University Press.

——— (1987) *The State of the World's Children 1987*. New York: Oxford University Press.

——— (1993a) *The Progress of Nations*. New York: UNICEF.

——— (1993b) *The State of the World's Children 1993*. New York: Oxford University Press.

——— (1996a) "Press Release, Secretary-General Reports Big Progress for Children." New York: UNICEF CF/DOC/PR/1996–24.

——— (1996b) *The Progress of Nations*. New York: UNICEF.

——— (2001) "Press Release, Polio Eradication: Final 1% Poses Greatest Challenge." Available online at http://www.unicef.org/newsline/01pr30.htm (April 3).

UNPD (United Nations Population Division) (1999a) *World Population Prospects: The 1998 Revision*. Vol. 1, *Comprehensive Tables*. New York: United Nations.

——— (1999b) *World Population Prospects: The 1998 Revision*. Vol. 2, *Sex and Age*. New York: United Nations.

——— (2001a) *Population, Environment and Development: The Concise Report*. New York: United Nations.

——— (2001b) *World Population Prospects: The 2000 Revision*. New York: United Nations.

——— (2001c) *World Urbanization Prospects: The 1999 Revision*. New York: United Nations.

UNSC (United Nations Security Council) (1948) Resolution 801, May 29.

UNSD (United Nations Statistical Division) (2001) *The World's Women 2000: Trends and Statistics*. New York: United Nations.

USBC (United States Bureau of the Census, International Programs Center) (1994) *World Population Profile*. Washington, DC: Government Printing Office.

——— (1997) "Money Income in the United States: 1996." *Current Population Reports*. Washington, DC: USBC, pp. 60–197.

USCC and AN (United States Code Congressional and Administrative News) (1991) 101st Congress, Second Session. Legislative History Clean Air Act Amendments, January, No. 10D. Minneapolis/St Paul: West Publishing.

USCEA (U.S. Council of Economic Advisors) (2001) *Economic Indicators* (May): 27.

USDJ (U.S. Department of Justice) (2000) *1998 Statistical Yearbook of the Immigration and Naturalization Service*. Washington, DC: Government Printing Office.

USDOE (United States Department of Energy) (2001) "Oil Dependence and Energy Security." Available online at www.fueleconomy.gov/feg/oildep.shtml (October 21).

USG (United States Government) (1995) *World Military Expenditures and Arms Transfers, 1993–94*. Washington, DC: Government Printing Office.

USGAO (United States General Accounting Office) (1991) "Child Labor: Characteristics of Working Children." Washington, DC: USGAO.

USSCEPW (United States Senate Committee on Environmental and Public Works) (1993) "Three Years Later: Report Card on the 1990 Clean Air Act Amendments, November 15." Washington, DC: Government Printing Office.

Valente, C. M., and W. D. Valente (1995) *Introduction to Environmental Law and Policy: Protecting the Environment Through Law*. Minneapolis/St. Paul: West Publishing.

Võ, X. H. (1994) *Oil, the Persian Gulf States, and the United States*. Westport, CT: Praeger.

von Laue, Theodore H. (1987) *The World Revolution of Westernization: The Twentieth Century in Global Perspective*. New York: Oxford University Press.

Warner, Sir Frederick (1991) "The Environmental Consequences of the Gulf War," *Environment* 33, no. 5.

WBCSD (World Business Council for Sustainable Development) (1996) *The Changing Future for Paper.* Geneva: WBCSD.

WCED (World Commission on Environment and Development) (1987a) *Our Common Future.* Oxford: Oxford University Press.

———— (1987b) *Environmental Protection and Sustainable Development: Legal Principles and Recommendations.* London: Graham and Trotman.

———— (1993) *The 1993 Information Please Environmental Almanac.* Boston: Houghton Mifflin.

Weber, Eugen (1964) *Varieties of Fascism: Doctrines of Revolution in the Twentieth Century.* Princeton, NJ: Van Nostrand.

Weber, P. (1993) "Reviving Coral Reefs." In Lester R. Brown, ed., *State of the World.* New York: W. W. Norton.

Weeks, John R. (1996) *Population: An Introduction to Concepts and Issues.* Sixth edition. Belmont, CA: Wadsworth.

————. (1999) *Population: An Introduction to Concepts and Issues.* Seventh edition. Belmont, CA: Wadsworth.

Weston, Burns H. (1992) "Human Rights." In Richard Pierre Claude and Burns H. Weston, eds., *Human Rights in the World Community.* Philadelphia: University of Pennsylvania Press.

WFA (World Federalist Association) (1996) *The Global Economy.* Part 2: *TNCs and Global Governance.* Washington, DC: WFA.

Wilford, Rick, and Robert L. Miller, eds. (1998) *Women, Ethnicity and Nationalism: The Politics of Transition.* New York: Routledge.

Wilkinson, R. G. (1992) "Income Distribution and Life Expectancy," *British Medical Journal* 304: 165–168.

Wiseberg, Laurie S. (1992) "Human Rights Nongovernmental Organizations." In Richard Pierre Claude and Burns H. Weston, eds., *Human Rights in the World Community.* Philadelphia: University of Pennsylvania Press.

Wiseman, Henry (1983) "United Nations Peacekeeping: An Historical Overview." In Henry Wiseman, ed., *Peacekeeping, Appraisals and Proposals.* New York: Pergamon Press.

World Bank (1993) *World Development Report.* Oxford: Oxford University Press.

———— (1994) *Investing in Infrastructure: World Development Report 1994.* New York: Oxford University Press.

———— (1995) *World Debt Tables.* Washington, DC: World Bank.

———— (1996) *From Plan to Market: World Development Report 1996.* New York: Oxford University Press.

———— (1997a) *The State in a Changing World: World Development Report 1997.* New York: Oxford University Press.

———— (1997b) *World Development Indicators.* New York: Oxford University Press.

———— (1999) *World Development Indicators 1999.* Washington, DC: World Bank.

———— (2000) *World Development Report 2000/2001: Attacking Poverty.* New York: Oxford University Press.

———— (2001a) *Global Development Finance.* Washington, DC: World Bank.

———— (2001b) *World Development Indicators 2001.* New York: Oxford University Press.

———— (2001c) *Attacking Poverty: World Development Report 2000/2001.* New York: Oxford University Press.

World Forestry (n.d.) "South American Forest Products Trade." Available online at www.worldforestry.org/wfi/trade_sa.htm (September 26, 2001).

WRI (World Resources Institute) (1992) *Global Biodiversity Strategy.* Washington, DC: World Resource Institute.

———— (1993) *Biodiversity Prospecting*. Baltimore: World Resources Institute.
———— (1994) *World Resources 1994–95: A Guide to the Global Environment*. New York: Oxford University Press.
WWF (World Wildlife Fund) (1997) "Rio+5: Earth Summit II Backslides on 1992 Promises." Press release dated June 27. Available online at http://www.panda. org/news/press/archive/news_134.htm.
Zhu, B., et al. (1996) "Cigarette Smoking and Its Risk Factors Among Elementary School Students of Beijing," *American Journal of Public Health* 86, no. 3.

THE CONTRIBUTORS

Elise Boulding is professor emerita of sociology at Dartmouth College and former secretary-general of the International Peace Research Association. She has undertaken numerous transnational and comparative cross-national studies on conflict and peace, development, and women in society. A scholar-activist, she was international chair of the Women's International League for Peace and Freedom in the late 1960s. Among her many publications are *Building Peace in the Middle East: Challenges for States and Civil Society* (ed.) (1994); *The Future: Images and Processes,* with Kenneth Boulding (1995); *Women in the Twentieth Century World* (1997); and *Cultures of Peace: The Hidden Side of History* (2000).

John K. Cox is associate professor of history and chair of the department at Wheeling Jesuit University, where he teaches courses on nineteenth- and twentieth-century Central and East European history and culture. He is author of *The History of Serbia* (forthcoming) and several articles on fascism, Marxism, and undergraduate pedagogy. He contributed essays on many European countries, including Poland, Slovenia, Hungary, Spain, and Greece to *Global Studies: Europe,* seventh edition (author/editor E. Gene Frankland, 2002).

Jennifer Dye is a student at Wilmington College, Ohio, double majoring in social and political science and communication arts. She has been the service site coordinator for Homelessness and Hunger, as well as working with Christians for Social Action, a Christian social service group that works toward social justice. Currently, she is researching the gender roles of women in society, specifically how women are viewed in pop culture.

323

George Kent is professor of political science at the University of Hawaii at Manoa. He has written several books, including *The Political Economy of Hunger: The Silent Holocaust* (1984); *Fish, Food, and Hunger: The Potential of Fisheries for Alleviating Malnutrition* (1987); and *The Politics of Children's Survival and Children in the International Political Economy* (1991). He is coconvener of the Commission on International Human Rights of the International Peace Research Association and coordinator of the Global Task Force on Children's Nutrition Rights. He has worked as a consultant with the Food and Agriculture Organization of the United Nations, the United Nations Children's Fund, and several nongovernmental organizations.

Ellen Percy Kraly is professor in the Department of Geography at Colgate University. She is author of numerous articles on international migration to and from the United States and settlement patterns within the United States, U.S. immigration policy and environmental issues, and trends in socioeconomic mobility among immigrant groups. She has conducted research for the United Nations Statistical Commission, the U.S. Immigration and Naturalization Service, the National Academy of Sciences, and the U.S. Commission on Immigration Reform. She has been president of the Population Specialty Group of the Association of American Geographers.

Jeffrey S. Lantis is associate professor of political science and chair of the international relations program at The College of Wooster. His teaching and research interests include foreign policy decisionmaking in democratic states, international cooperation and conflict, and European politics. He is author of *Domestic Constraints and the Breakdown of International Agreements* (1997), coeditor of *The New International Studies Classroom: Active Teaching, Active Learning* (2000), and coeditor of *Foreign Policy in Comparative Perspective: Domestic and International Influences on State Behavior* (2002).

Marian A. L. Miller is associate professor in the Department of Political Science at the University of Akron. Her major areas of research interest are environmental politics and the politics of development. Her book, *The Third World in Global Environmental Politics*, received the International Studies Association's 1996 Sprout Award for its contribution to international environmental politics. She has also authored numerous papers on various aspects of environmental politics.

Bruce E. Moon is professor in the Department of International Relations at Lehigh University. He is author of *The Political Economy of Basic Human Needs* (1991), and *Dilemmas of International Trade*, second edition (2000).

His articles have appeared in *International Studies Quarterly, International Organization*, the *American Journal of Political Science, Comparative Political Studies, Political Research Quarterly,* and the *Journal of Conflict Resolution*. His research in international political economy and foreign policy has also appeared in several edited volumes.

Marjorie E. Nelson is an emerita faculty member of the Department of Social Medicine at the Ohio University College of Osteopathic Medicine. After residency training in Philadelphia, she was staff physician with the American Friends Service Committee Rehabilitation Project at Quang Ngai Hospital in Vietnam from 1967 to 1969. She has been a local health officer and medical director of a Planned Parenthood affiliate, and she has worked with the Hospital Ship HOPE in Guinea, West Africa.

Don Reeves, now in semiretirement on his Nebraska farm, served as interim general secretary for the American Friends Service Committee during most of 2000. From 1977 to 1980, he was legislative secretary for the Friends Committee on National Legislation, the Quaker lobby group in Washington, D.C. In 1983, he was a founding chair of the Nebraska Farm Crisis Response Program under Interchurch Ministries of Nebraska and, earlier, of Nebraskans for Peace. He served as economic policy analyst on Bread for the World (BFW) Institute staff from 1987 to 1998 and continues to consult with BFW Institute and other concerned citizens on hunger- and poverty-related issues. During the early 1990s, he directed a church- and farm-sponsored educational effort on issues of U.S. agriculture, trade, and development.

Gerald W. Sazama is associate professor of economics at the University of Connecticut. He was a Fulbright fellow for research in energy economics in Costa Rica and held a Social Science Research Council fellowship to research land taxation in Chile. He was also a consultant on project evaluation and taxation for USAID in Bolivia, Costa Rica, and Nicaragua and has done consultation work with the World Bank on a training project in Afghanistan. He has published in the *National Tax Journal*, the *Journal of Regional Science, Europe-Asia Studies,* and *Economic Development and Cultural Change*.

Karrin Scapple is on faculty at Ozarks Technical Community College in Springfield, Missouri. Her research interests include international environmental politics and policy, the United Nations, and international law. She has published in the *Journal of Environment & Development*, the *Journal of Environment & Security, International Studies Notes*, the *American Society of International Law Antarctic Interest Group Newsletter,* and several edited volumes.

Mark Seis is assistant professor of sociology at Fort Lewis College in Durango, Colorado. His primary research interests are environmental crime, law, and policy. He has published on various topics ranging from economic globalization and the environment to Native American perspectives of environmental crime to ecological problems with the Clean Air Act. He is also coauthor of *A Primer in the Psychology of Crime*.

D. Neil Snarr is professor of sociology and director of international education at Wilmington College in Ohio. He also directs the Global Issues program, a three-hour course required of all freshmen and seniors. He has published in major journals and edited several books.

Michael T. Snarr is assistant professor of social and political studies at Wilmington College in Ohio. He is also coeditor of *Foreign Policy in Comparative Perspective: Domestic and International Influences on State Behavior* (2002). His research focuses on Latin American foreign policy, and he teaches courses on global issues, global politics, U.S. foreign policy, and the United Nations.

Carolyn M. Stephenson is associate professor of political science at the University of Hawaii at Manoa. She is editor of *Alternative Methods for International Security*. She was director of peace studies at Colgate University and was coeditor of *Peace and Change: A Journal of Peace Research* for a number of years. The author of articles on peace studies and conflict resolution, she is currently researching nongovernmental women's, environmental, and disarmament organizations at United Nations conferences.

INDEX

ABOUT THE BOOK

Reflecting the enormous changes—and challenges—of the five years since the first edition of *Introducing Global Issues* appeared, this fully revised text explores the various dimensions of conflict and security, the global economy, development, and the environment.

The material is designed to be easily accessible to readers with little or no prior knowledge of the topics covered. Each chapter provides an analytical overview of the issue addressed, identifies the central actors and perspectives, and outlines past progress and future prospects. The book is enriched by challenging discussion questions posed to enhance students' appreciation of the complexities involved as well as by suggestions for further reading.

Michael T. Snarr is assistant professor of social and political studies and **D. Neil Snarr** is professor of sociology, both at Wilmington College.